Reinventing Physics:  Logic and Physics

A Dialectical Approach to Physics

A Thesis
Submitted to the Faculty of the Graduate School
of the Hellenist America Institute--University

by

Patrick Aquinas O'Dougherty

in Partial Fulfillment of the Requirements
for the Degree of
Doctor of Philosophy

September  1993

The Hellenist America Institute Publishing Company
Riverside Plaza  M3410
1615 South Fourth Street
Minneapolis, Minnesota  55454

O'Dougherty, Patrick Aquinas, 1946--

    Logic, Dialectics and Physics
    Includes Index and Bibliography

### Dedication

To Richard Kast and Patricia O'Dougherty-Kast.  To my father, James O'Dougherty, my brother Mike, and my sisters:  Margaret, Mary Ann, and Maureen.  To Megan, Sean, and Stephen.  To David Noble, John O'Dougherty, John Howe, Mark McGee, John Herlick, the Conant family and especially to Mike Franey and my namesake St. Thomas Aquinas.

ISBN:  0-9626665-2-1
Library of Congress Catalog Card Number:  93-61179

Printed in the United States of America

# Table of Contents

## Reinventing Physics: Introduction

The aim of this book is to try to reinvent the field of physics. The key to the history of physics, relativity, small particle physics, and unified field theory is logic. The covert truth of physics is logic. A specific approach to logic developed here is dialectics. Intellectual history and physics are equivalent fields. Plato, German idealism, specifically Hegel, are significant to the framework of physics. A biblical revelation is germane to creation: "In the beginning was the word." Thus, idealism has a biblical basis. A conceptual framework from Hegel is used to add a new dimension to physics. An example of a dialectical approach to the universe is a quote from David Noble. He thinks "birth, death, and rebirth" are the basis of the universe.[1] Thus, there is thesis, antithesis, and synthesis in nature. The basis of physics is life, death and reinvention. Hegel's idea of the Spirit hits at a loftier level of meaning than the individual fact or experiment.

Little work has been done to apply dialectics, especially Hegel's philosophy to physics. To give focus to the idea that logic is the covert basis of physics the writer is going to build a conceptual framework from Hegel's philosophy of physics. Then Hegel and science, a dialectical approach to science, Hegel's logic, and a modern approach to logic will be presented. Finally, This conceptual framework will be applied to the history of physics and an oriental mystic approach to physics. Ideas are born, die, and are reborn. It is time for Hegel's rebirth. In the second part of this book the spirit of physics (Is it an oxymoron?) will be captured by studying the history of the physics department at the University of Minnesota.

Hegel thinks that truth is a triad. Is there any sign that this is correct. The commutative, associative, and distributive laws of addition and multiplication form a triad. It is not a dialectical triad but it is an elementary triad. Moreover, in mathematics there is the inverse, and there are inverse functions. The antithesis of the dialectic is like an inverse. The theme of synthesis is critical to many areas of thought. Also, one can turn a number into an absolute number. The Spirit or the Absolute hits at a higher level than the physical. God is beyond abstraction. The Trinity is a triad. The idea of antithesis should be given a broad meaning like deviation, or aberration or development or inverse or anomaly or approximate function.

## An Example of a Triad in Mathematics and Physics

"Commutative Property of Addition: $a + b = b + a$"

1

## Logic, Dialectic and Physics

"Associative Property of Addition: $(a+b)+c = a+(b+c)$"
"Commutative Property of Multiplication: a times $b = b$ times    a."
"Associative Property of Multiplication: (a times b) times c =    a times (b times c)."
"Distributive Property of Multiplication over addition: a    times $(b+c) = a$ times b + a times c."[2]

### The Inverse

"Principle of Inverse of Variation: y varies inversely as x    means $y = k$ divided by x for some constant k."[3]

An inverse can be taken of many numbers and equations. The law of gravity and the law of magnetism have an inverse relation. Thus, there is much relevance of Hegel to physics. Some thinkers have argued that mathematics is essentially dialectic. This is true only in part.

### A Background on Hegel from Philosophy

Is the universe of physics due to chance or planned? Is the mind special or merely plumbing? Is matter an illusion?[4] Hegel suggests that the universe is planned, the mind is special, and mind and matter are a unity.

Hegel's philosophy is monistic idealism. It has a negative and a positive side. The negative part criticizes philosophies which contend that reality is unknown or that it contains several particular truths. The positive part of Hegel's philosophy suggests that reality is the Person or Spirit.[5]

Hegel argues that the human _mind_ cannot contain all truth. Thus, no human system is completely true. All human systems have errors and the mind looks towards ideas that refute or qualify a system. Thus, truth is a dialectical process or the combination of two partially true ideas united in a synthesis or "triad."[6] God's revelation does not have errors. Consciousness and gravity are universal truths. The _human_ mind has limits.

What sort of process is this dialectic? Hegel suggests that one cannot analyze Pure Being. It is really Nothing. There is a paradox here: Since one can think about Nothing, it is not actually Nothing. Thus, Pure Being or Nothing point towards a Being with determined characteristics.[7]

### The Universe and the Self and the Not Self

## Logic, Dialectic and Physics

Most people distinguish the Self from the rest of the universe and from the Not Self. One perceives the Not Self but is different from it. Hegel suggests that the Self knows itself but it separates itself from the Not Self. He synthesizes the Self and the Not Self in cognition. Reality forms a triad. Hegel argues that absolute reality is knowable and it is not featureless.[8]

What is Hegel's positive doctrine? It is the idea that ultimate Reality is "the absolute." It is a principle of logic which expresses itself as order in the universe. It is all of existence and thought. It contains space and time and the process of history and thought. History is the process or revelation of this principle's nature. It is a revelation which forms an experience in limited minds.[9]

The mind is self-conscious. Thus, the absolute is self-conscious. It is not incompatible with the idea of God. One finds the attributes of God in religion. There is a unity of God and Being. The universe is a unity.[10]

### Internal Relations

Change something and you change everything in the world. A clock on the wall changes every time someone shots a lion in the jungle. This is a conclusion of an idea that the universe is an interrelated unity and a whole. Articles exist in relation to other items and these items like the relation of a pigeon's egg to a hen's egg are part of their nature. A whole is more than a mathematical sum of parts. A picture is a whole, for example. A "concrete universal" which contains the abstract and the sense experience is the real. This is Hegel's truth.[11] How does the big bang theory of the universe explain logic? It is logic. God is the unity of the universe.

### An Outgrowth of Hegelianism: Dialectical Materialism

Karl Marx (1818-1883) and Friedrich Engels (1820-1895) pronounced dialectical materialism as opposed to idealistic dialectics. Marx accepted Hegel's idea of the dialectic. He applied it to objects and to history. Scientific truth contains its opposite. Marx was not completely deterministic. He thought that human nature changes. Marx contended that matter came before mind in the universe. Idealists, like Hegel, maintain that the mind came first.[12]

Hegel thought that an object "is a focal point for a set of internal relations." Marx, in contrast, focuses on the individual as a set "of social relationships." He adapts Hegel's word "Dialectical" to his philosophy of materialism. Marx then applies this philosophy to economics and history.[13]

3

Logic, Dialectic and Physics

## Scientific Materialism:  The Field of Physics

Materialism is the main outlook of scientists.[14]  However, today the field of physics has challenged the tenets of materialism.  Matter is seen as being mysterious and weakened.  Symbols and mathematical equations are the basis of the new physics used to describe the electron or the pion.[15]

Logic, Dialectic and Physics

## Hegel and Physics: A Perspective from Eduard Farber

Dialectics developed out of Greek philosophy. Hegel invented the word "dialectic." It means a process between thesis, antithesis, and synthesis. Dialectics are a logical process.[1] The basis of physics is logic. Mike Franey thinks "existence is logic."[2] The big bang that created the universe was a product of mind and logic. Creation is logic.

In Hegel's first book on logic, <u>Jenenser Logik</u>, Hegel presents space, time, and force as objects that waver and change. They go over and become their opposites. They move from the real to the negative. How are they controlled? Hegel does not answer this question. Objects move from the necessary to the free. In his writings Hegel looks at Act and Process. Thus, cognition is "the process of discerning." Objects shift in consciousness or undergo "inversion." Hegel looks forward to process theology.[3] The idea of a set would appeal to him. However, are the algebraic rules of commutative, associative, distributive and inversion laws of mathematics dialectic? The answer is only in part.

Hegel thinks that cognition directs itself to the other. Reflection is a relation that moves, changes, and disappears. Dialectics comes out of a reflection of the world which is a dialectic. The world contains "complementary pairs of opposites." Existence changes. Change is becoming an opposite, for example, a pendulum. There is kinetic and potential energy. There is matter and anti-matter.[4]

### Infinity

Hegel contends that dialectics have a "specified content." And thoughts, many of them, are concrete. Some of Hegel's thoughts are similar to the first law of thermodynamics. Also, the equation of the positive and the negative relates to the law of conservation of matter and energy. Hegel thinks that a quantum particle exists "independent of magnitude." Some elements in calculus exist only in relation to each other. **Logical and mathematical relationships extend to objects**. Facts are insignificant unless they show relations. A substance and its characteristics relate in a "reversible relationship." Relationships are interesting if they are a "feature of facts." Theories highlight features. For example, the attraction of opposite poles and the hostility of like poles show the essence of magnetism.[5] Infinity is a part of the ordinary. Modern physics has many absolutes. Are we moving towards an absolute in physics? Perhaps we are. What is the opposite of the absolute? Maybe it is as Mike Franey thinks the irrelevant or the absurd.[6] Maybe

5

dialectics are the absolute. Is the universe finite or infinite? The answer is we do not know.

## A Definition: Force

The main law of mechanics, force = mass x acceleration, shows how forces move masses. It does not tell us what a force is. Physics does this. In a way, force has a dual relation. It is a being within itself and a relativity to others.[7]

## The Fluid

Hegel thought that matter is an "absolute fluid." Heat is a fluid. Being and nothingness unite in Becoming. Reality is a process. Thus, Hegel had a philosophy somewhat similar to the functionalist position. Force, cause, and element play roles in a process or functionalism. Man is an experimenter. He is like God. One discovers nature by the idea. Man is an actor in operationism. Operationism is a universal word referred to the "free activity of Idea." Thus operationism is part of the universal process. It is little operations created by Idea.[8]

## Hegel's Ideas and Experimental Truth

Hegel thought that existence is changing. He focuses on change as inversion. The arrival of Being relates towards falling away. We separate reflection into parts. For example, "trembling-in-itself" can be broken into parts. Also, in spectroscopic analysis we must understand that in the atom there is a relation that is made by vibrations. Hegel unites Spirit, notions, ideas, subjects and experiments by inversion. Hegel did not predict electrons, but if discovered he could have foreseen the positron. Hegel contends "the Absolute is Spirit."[9] Is there an anti-Absolute? What about zero? From the perspective of modern physics, matter and energy are complex and subtle.

Hegel does not accept "force and affinity" when they exist apart from relations between mass and movement. Matter has a "trembling-in-itself." The Yin and the Yang come together. Spirit is the truth."[10] Einstein equates potential and kinetic energy on the earth with all the energy in the universe. He synthesizes potential and kinetic energy.

Logic, Dialectic and Physics

## Hegel and Science:  Paul Thagard

Much research in recent years in the philosophy of science has focused on the development and structure of knowledge or "historical epistemology." Paul Thagard claims that Hegel was an "historical epistemologist." His dialectics, for example, deal with the development of knowledge. Hegel relates dialectics to consciousness, to history, to logic and to philosophy.  Dialectics apply to science.[1]

Hegel analyzes the development of consciousness from the primitive idea, Being, to the Absolute Idea.  Thagard argues that Hegel's dialectics and the expansion of scientific knowledge are clearer in the dialectic stages of consciousness than in the development of the concept.   Hegel's Phenomenology deals with the "growth" of ideas by subjects.[2]

Hegel calls a succeeding stage in the dialectical process its negation.  A negation is often complex.  A new level supersedes and contains each stage it negates.  There is a clearing away and a preservation.  What is the anti-Absolute?  It is the "irrelevant."  Thomas S. Kuhn argues that scientific breakthroughs are not mainly an accumulation but more like a dialectical negation.  In contrast to positivists who think that a theory is a group of sentences, Sneed and Stegmuller relate a theory to set theory.  Hegel was unfamiliar with set theory, but, he favored the idea that "the representation of knowledge is essentially nonsentential."  For example, in cognitive psychology frames have supplemented "sentential representation of knowledge."  Frames contain more data.[3]  A frame in a computer program might be an example.

Another example is Newton's optical theory that particles compose light. Young and Fresnel replaced Newton's theory with a wave theory of light which helps explain polarization and diffraction.  Planck and Einstein negated the wave theory with a photon theory.[4]  Is the Absolute a convergence of theories?  Perhaps, it is.  Is the absolute a constant?  Hegel thinks that it is Spirit.  Present day "process theologians" think that we are evolving with God. They think there is an evolutionary dimension to God.  Religion is a "process." Spirit is a process.

### "Beyond Sentences"

What are the differences between set theory and Hegel's dialectic? First, Hegel's dialectic contains previous ideas.  Second, the dialectical stage is a product of earlier stages.  Third, there are many empirical contradictions.[5]

7

**Logic, Dialectic and Physics**

How is Hegel's dialectic similar to the evolution of scientific knowledge? First, it is a dynamic model. Second, the model represents more than an accumulation. For example, dialectics contain preservation and cancellation. Finally, knowledge is not just sentences, but, contains many complex structures. There are formal proofs for the dialectic.[6] How does Hegel explain irreversible chemical reactions? He does not. Hegel does not deal with the field of physics often.

Logic, Dialectic and Physics

## Dialectics and Modern Science

Errol Harris is a realist. He thinks that the philosophy of "idealism" is largely a defense of the position that people know objective reality through reflection. The rational is an evolving set of judgments within the context of science. Harris thinks that empirical theory justifies rationality. A priori logical forms are largely mythical.[1]

Harris finds rumors, like Nietzsche's, of God's death are fallacious. He thinks that twentieth-century empirical theory and research have favored a holistic evolutionary nature that cuts across many scientific endeavors. The Logical Positivists fail to explain holism. They are wrong in their reduction of philosophy to the philosophy of science. They are also wrong in writing off the contributions of speculative philosophy.[2]

Moreover, Harris emphasizes the incompleteness of atomism and reductionism in contemporary scientific interpretations. He finds teleology critical to scientific interpretation. Today one finds in physics the "anthropic cosmological principle" "as a necessary principle" in cosmological theory. John Barrow's and Frank Tipler's The Anthropic Cosmological Principle unites different versions of the anthropic cosmological principle with teleology.[3] Anthropic means man or human centered.

Besides criticizing Carnap's and Ayer's logical positivism, and Wittgenstein's and Russell's 'analysis', Harris contends evolution and dialectical reason are also pertinent explanations to the universe. Hegel maintained that formal logic and not the dialectic is a good mode of reason for the empirical sciences. Nature, in contrast to science, is a proper area of philosophical analysis. Empirical research today supports the idea of a hierarchy of connected forms of greater complexity, mainly from the inorganic to the psychological and social. This "holistic drive" is behind much scientific progress in the twentieth century. This unified vision of natural science is a product of monotheism.[4]

Harris thinks the logical and ontological portends of empirical research are not completely distinct. Dialectical onto-logic is the framework of Harris's system. Hegel had a holistic metaphysics which contains the idea of the dialectic. In Harris's endeavor, logic is progressive and there is a dialectic. The part relates to the whole. In his Hypothesis and Perception, Harris criticizes Empiricist contributions to logic and scientific research and suggests a dialectical approach.[5]

## Logic, Dialectic and Physics

### Errol Harris and the Problem of Internal Relations

What is the problem of internal relations? An example of this problem would be if A relates to B, then the absence of that relation leaves A the same as it was, therefore the relation is external. If after the relation is absent A is different, then the relation is internal. For example, if a train moves from New York to Pennsylvania is it still the same train? The answer is yes and the relations are external. Again, if A and B are the numbers two and three the relationship is "indispensable and internal."[6] Physics research contains both internal and external relationships.

"Logical atomists" like Wittgenstein and Russell argue that the world contains a pile of unrelated facts. Hegel, in contrast, holds the world to be absolute with the nature of the parts dominated by the whole. Harris favors this approach. This holism applies to relativity and quantum mechanics. Einstein contends that every body has a relation with all the other bodies in a "four-dimensional space-time whole." Hegel has an idealistic monism. Furthermore, Harris contends that on all the main evolutionary levels the gestalt pervades. For example, flatworms, when cut up, can grow new heads and tails.[7]

Evolution is holistic. The intellect analyzes and synthesizes wholes and parts. Consciousness and feelings are a level of unity. Harris thinks that the nervous system integrates our senses and feelings. The mind is a "set of impulsions." Philosophy is a continuation of science.[8]

What are the signs of internal relations? First, they overlap with the whole like spokes do to each other in a wheel. Second, relations embed themselves in a continuous matrix. Third, about terms in a continuum, there are differences of degree like colors on a "color continuum." Fourth, logic cannot deal adequately with actual thinking which is much more complex than telephone circuits.[9]

Harris thinks that consciousness is the brain's matter organized and integrated in a particular way. "The organism is a whole of wholes." How does one integrate unified wholes into the absolute? If the absolute is personal, how can one person exist in another? How can a "perfectly integrated world" contain "imperfectly integrated wholes?" Harris replies by saying that it is his thesis "consciousness-or rather, feeling, which is the matrix of consciousness-is the _form_ of the organization, at a specially high level of integration, of physiological activity."[10]

## Logic, Dialectic and Physics

What is "matter?" Today many physicists think that it is a "pattern or Gestalt of energy,"--"a 'wave packet,' which is an elementary particle." The nucleus of an atom contains organizations of these particles into complex force fields. Consciousness organizes matter into a complex "living activity." Feeling is the form of organization of consciousness.[11]

Harris defends holism by noting that experiences belong to the single subject. Do personalities overlap and have difficulty fitting into the Absolute? Harris cites the example of lovers, children and the Trinity.[12]

Can imperfectly integrated wholes make up a perfectly integrated world? Harris points out that an organism evolves from a germ-cell to fulfill in the adult. The absolute whole can include "lesser wholes."[13]

### Teleology and Science

Can a future event causally influence a prior event? Can purposive explanations be correct only for cases of conscious behavior? Harris thinks that it is not certain that the end activity of a process has any relationship to the starting aim. He thinks that conscious relationships between initial aims and final ends are largely accidental. Do self-fulfilling prophecies exist in science? Harris argues that a teleological explanation understands parts in terms of the whole. The determining factor is implicit in every phase of the process. Thus, physics and biology are holistic in structure and understanding.[14]

In the evolution of an organic series successive entities "sum up" preceding forms; and, the organic sum shows itself through a dialectical generation of successive forms. Theism is a necessary point of view about teleology. There is a "chain of being." George Lucas gives an example of this chain. Epidermal cells are small biological entities and also parts of the entire body. Work in "nonlinear dissipative structures evolving order from randomness" supports Harris's teleology in the natural sciences. Teleology is good theology. Hegel's implication for physics is physics is teleological.[15]

Robin G. Collingwood develops a "scale of forms" where successive entities sum up features of earlier forms. Thus, building an "organic totality." This sum of sophisticated forms reveals itself through a dialectical production of successive forms on the scale. Human forms are at the top of "the human scale of forms." The implication is that evolution is dialectical. The end of evolution is transcendent, whole and self-realizing.[16]

## Logic, Dialectic and Physics

About teleology Harris argues that the universe is a "whole" and is a dialectally structured hierarchy which directs itself towards a consummation. This involves a principle of order that is transcendent. Life directs itself towards more life. The idea of forms points towards an infinity that Spinoza called "infinite Substance." Hegel termed this "absolute Spirit" and Anselm adored this as God.[17] Logic is evolving and Hegel's dialectic has a dialectic or approximate function aspect to it. There might be, however, a reductionistic truth to complexity--dialectics. For example, there are a limited number of elements in physics and chemistry.

## Harris's Commentary on Hegel's Logic

What is the relationship in logic between form and content? Russell and Wittgenstein argue for a "neutral" logic. Hegel, in contrast, relates logic at all levels to content. He locates a system of categories in the content itself. Harris argues that logic is not something neutral about items and postulates--an object like "atomic particulars" that relate.. In contrast, Russell and Wittgenstein argue that atomic facts have an 'logical atomism' which may or may not relate to the rest. Hegel did not reject formal logic.[18]

Hegel made a distinction between understanding and reason. Understanding means abstraction and distinctions. Hegel thinks that understanding does not show internal relations between the distinctions. Understanding needs negation--an object is one and "not another thing." Reason, in contrast, synthesizes a concept which is a principle of order and not simply an abstraction. Unified field theory is an example of this concept. Is arithmetic dialectical? Engels thought this. It is in part. The dialectic can erupt in history.[19]

Should Hegel have started with nothing instead of being? Hegel presents Being-Nothing-Becoming as the "beginning." For the Empiricists, the smallest unit is the term, and for Kant it is the judgment, for Hegel it is the system. Harris cites the example of DNA in the genetic code as a "self-specifying whole."[20]

## The Unity of Thought and Being

Hegel thinks that thought is wholeness. In contrast, Aristotle supports the proposal there are ideas and objective forms of thought. Kant has the notion of active forms of "thinking" as a transcendental part of experience. Hegel and Aristotle think that logic is the science of thought separate from "empirical-real thought." Hegel searches for "synthetic" conditions in

12

## Logic, Dialectic and Physics

experience. His idea of "concept" is similar to the transcendental Ego. Harris suggests, in contrast to Aristotle and Kant, that Hegel argues for the "totality and infinity of pure thinking."[21]

Hegel's thought is holistic because it relates the logical form to a rational principle. This dialectic leads to the realized concept, the Absolute Idea. Thought is an "infinite systematic totality." The Infinite whole sublates or denies the finite.[22]

Hegel argues "Thought and Being are an Identity." The universe is a "product of mind." Identity is an evolutionary process.[23]

### Harris's Reflections on Hegel's Logic

Thomas Rockmore argues that philosophers should not lose track of the history of philosophy. Hegel's systematic position needs the perspective of the previous history of philosophy. Harris tries to define Hegel's position as a German idealist. Rockmore suggests the history of metaphysics from Parmenides to Kant contains several dialectically related strategies on the study of objectivity. Hegel argues speculative logic is a form of metaphysics, "the science of things captured in thought." Thus, the history of metaphysics is inseparable from critical philosophy. Hegel's system opposes and supersedes previous philosophies in a dialectical process. Thus, history relates to system. The history of physics relates to system.[24]

Logic, Dialectic and Physics

## A Portrait of Hegel

Hegel was a speculative idealist. Other exponents of this point of view are Fichte and Schelling. These thinkers held in contrast to Kant that metaphysics or cognition or the cognition of the universal is possible. Hegel thought that the universe is rational. Reality is reason: "Whatever is rational is real and whatever real is rational." The essence of the universe is reason. Logic and metaphysics are the same. Concepts lead to their opposites. Each negation leads to a new synthesis.[1] There is a logical progression in nature.

Hegel thought reality goes through three stages: "being in itself, being for itself, and being in and for itself." The stage of "Being in itself exists as potential." It is not complete. After developing a difference from "Other being" it becomes "being in and for itself." The Spirit becomes "other being" or nature developing in space and time.[2]

Consciousness is nature's negation. Consciousness contains both the "in and for itself of the spirit." Consciousness moves through three planes. It moves from the subjective to the objective. Then, it finds its apex "in absolute spirit." Philosophy has three divisions. The first is logic which is the discipline of the "in itself," the second is study of nature or the "for itself." The final stage is the philosophy of the Spirit. Anthropology, psychology and phenomenology are the three disciplines that deal with the spirit.[3]

Human reason or objective spirit deal with nature. Morality is an area of objective spirit. It has three part also: ethics, law, and history. The spirit represents art, religion, and philosophy. They are the reasons, apex, and standards for the area of the absolute spirit. This absolute spirit deals with the philosophy of art, religion, and the history of philosophy. Hegel maintains the dialectic throughout his philosophical purview. Reality has a rhythm.[4] Dialectics are like a dance. They are a rhythm.

Hegel contends that the ideal state is the reality of the ethical ideal. The divine enters into life through this stage. History contains reason. What is Hegel's philosophy of history? It is the idea that the World-Spirit acts through people in history. Reason has man's passions develop its ends. Nature contains the World-Spirit. Notions pass through the stages of youth, adulthood and death. History is evolutionary. The end of evolution is liberty. History is the process of the consciousness of liberty. Philosophy should try to comprehend progress.[5]

14

### Logic, Dialectic and Physics

The basic part of liberty is the spirit. The state is the chance of liberty. The law governs the individual. Christianity brought real liberty. The adolescence of western civilization was the Greek and Roman period where freedom arose. "Man" is free.[6]

Initially, Hegel favored the French Revolution as a step toward liberty. Later, Hegel favored a constitutional monarchy. Philosophy is a period understood in thought. He idealized the Prussian monarchy. Hegel felt that the old age of civilization would bring "wisdom and superiority."[7] David Noble thinks that history is "complexity."[8] The intellectual history of Hegel may be complexity. Physics is "anthropic."[9]

## Logic, Dialectic and Physics

### More on Hegel's Logic

Logic is the field of the pure idea. It is an abstraction. It is not just a method. It is a self-developing totality of [thought] its laws and peculiar terms." Logic deals with abstraction as opposed to sense perceptions. It works with quality, magnitude, potential, actuality, the one, many, 'Is and Is not.'[1]

The object of logic is thought. Senses deal with the individual. Logic focuses on metaphysics--the field of objects held in thought. The ideas and principles of logic deal with categories. The difference between subject and object typically disappears. Reason is the major principle of the world.[2]

Logic studies "pure thought-forms." The syllogism is a universal form of the universe. Consistency and truth are the problems of logic. Experience and reflection are two different methods of getting truth. These methods do not get absolute truth. Freedom characterizes pure forms of thought. Other forms of truth are "immediate knowledge," "trust, love, faith and religious experiences," and philosophical cognition. In Hegel's Phenomenology of the Spirit, he moves from "immediate consciousness" to the philosophical orientation. Process is a critical part of the philosophical outlook.[3]

### "Empiricism"

The basic idea behind empiricism is that what is true must be in the world and clear to sensation. This philosophy goes against the idea of 'ought to be.' Man must see and feel for himself and his presence in all facets of knowledge. Thought has no power except abstraction and identity. However, after using the categories of mass, force etc. it draws conclusions and in so doing uses the syllogistic form. Empiricism deals with experience or matter and form. Law and ethics are works of chance and lack "inner truth."[4]

The so-called Critical Philosophy assumes that experience is the basis for cognition. Experience is subjective. A priori elements are subjective. Categories lie in the 'I' in thought. Kant calls this the 'transcendental unity of self-consciousness.' 'I' is 'the transcendental unity of self-consciousness.' It is the 'Ego' Kant thinks that gives the marks of universality and need. The categories do not express "the absolute."[5]

Kant looks at an unconditioned entity, the Soul. He finds that within his consciousness 'I' and the fashioning subject. I am single, am the same, and am different from objects besides myself. Facts differ from intellectual formulation.

## Logic, Dialectic and Physics

Kant thinks God is reality or realities. Spinoza thinks that God unites thought and extension. Kant considers practical reason to show a <u>Thinking Will</u>. Judgment has a reflective dimension or power. The system of empiricism is also materialism and naturalism.[6]

## Immediate Knowledge

Thought is the faculty of finitisation. Hegel contends that immediate knowledge is like a fact. For example, in Descartes 'Cogito, ergo sum' or 'I think, therefore I am' there is immediate consciousness. It is part of truth.[7]

Jacobi and Descartes make three points. First, thought and the being of the person are one. Second, God unites with existence. Third, external objects are immediately a part of consciousness. Descartes moves from these unproved assertions to the whole scope of knowledge. In contrast, Jacobi suggests that cognition proceeds by finite mediations. Thus, we can have only an abstract idea that God is.[8]

## A Further Definition of Logic

Logic has three dimensions: the abstract, the dialectical, and the speculative. It can be further defined to the doctrines of Being, Essence, notion and idea. "Becoming is the unity of Being and Nothing." Quantity is the purity of being. The idea of number has two qualities--sum and unity.[9]

We should see "the infinite quantitative progression as only the meaningless repetition of the same contradiction which attaches to the quantum. What is the quantum?" It means how much or limited quantity. It is the first move towards a metaphysics which tries to comprehend the universe as number. Pythagoras thought that the essence of objects is number. What is the meaning of to "measure?" Measurement is a quantum to which a quality connects. In measurement quantity and quality unite.[10]

## Essence

Essence is Being reflecting upon itself. The Absolute is Being and the Absolute is Essence. Essence is Being gone into itself or the negation of the negative. Negativity is thus its own dialectic. Identity is reflection-into-self or abstraction. The absolute is that which is same with "itself." The absolute often means the 'abstract.'[11] Physics unites the abstract and the physical.

## Logic, Dialectic and Physics

Reflection contains the mark of difference, diversity or variety. Difference is positive or negative. Polarity in physics means opposition. Existence unites reflection into the self and reflection into the other. Kant's 'thing in itself' is the abstract. The concept of thing is the unity of existence meaning a coming together of identity and difference.[12]

What is matter? It is "Thinghood." Form is "the reflective category of difference." Matter is the unity of "existence with itself." Different matters coalesce together into one matter. Form is the category of difference in reflection. It is content. The whole and the parts correlate. "Force is a whole." Actuality is chance. Contingency is another facet of actuality. Actuality is also a unity. A substance is a sum of accidents. It is also cause. The truth of the necessary is freedom.[13]

### Notion

The Doctrine of Notion is the idea of freedom or the strength of a self-realized substance. It is development. The doctrine of notion has three parts (1) subjective notion, (2) objectivity, and (3) idea or absolute truth. Subjectivity contains the universal, the particular, and the individual. These terms are the abstract. The universal contains both the particular and the individual. The notion as the particular is the judgment or "the individual is the universal." "The subject is the predicate."[14]

### Syllogism

The Syllogism unites the notion and the judgment as one. It is notion, judgment and reason. There is the individual, the universal and the notion united in the abstract. In case of the rational Syllogism the subject is "coupled with itself." In the Qualitative Syllogism "a subject as Individual couples (concluded)J with a universal character by a (Particular) quality."[15]

What is the Mathematical Syllogism? It is the idea "if two things are equal to a third, they are equal to one another." What is the Syllogism of Allness? It bases itself on induction which bases itself on analogy. For example is the statement 'all metals conduct electricity.' There is an induction here: Gold is a metal so gold conducts electricity.[16]

What is the object? It is "immediate being." "What is the mechanism?" It is a unity of differences. There is the "Mechanism with Affinity" such as gravity or absolute mechanism such as the "non-independence of the objects." An example is the relative center.[17]

## Logic, Dialectic and Physics

There is the I-P-U syllogism, the U-I-P syllogism and the P-U-I syllogism. I means the individual. P means particular, and, U means universal. What is the end? It is a "contradiction of its self-identity against the negation stated in it." What are the means? They are objectivity made servant to purpose. What is the Idea? It is truth in and for itself. The Absolute is the Idea. The Idea is reason. It is process. What is Life? It is the immediate idea. Analysis and synthesis are two different forms of scientific method. What is Will? When subjective ideas get an original and objective determination they become Will. What is the Absolute Idea? It is the unity of "the Subjective and Objective Idea." It is the thinking Idea or "Logical Idea." We now have the "Idea as Being." Nature is the "Idea which has Being."[18] Physics is the Idea that has "Being."

Logic, Dialectic and Physics

## A Background in Logic:  A Textbook of Logic

There are two major divisions of Logic:  Formal Logic and Inductive Logic.  Formal Logic deals with inferences drawn from propositions, for example, immediate inference, involving a single proposition, or mediate inference, involving a variety of propositions.  Inductive logic studies inference of several kinds which are drawn from facts to help explain these facts, for example, analogy, circumstantial evidence and chance.[1]

A proposition involves actual judgment and potential judgment.  It is an abstraction.  One proposition infers from another.  Subjects and predicates are the terms of the proposition.  Categorical propositions deny or affirm a predicate of a subject, for example, "the proposition S is not P is true."  There are four types of categorical propositions.  They are, for example, (1) universal affirmative, (2) particular affirmative, (3) universal negative, and (4) particular negative.  The words **every** and **some** are ways to distinguish between the universals and the particulars, for example, "Some S's are not P."[2]

Rule: "No term that is undistributed in the premises, may be distributed in the conclusion, unless the laws of thought warrant it."  "If a given term is not distributed in the premises, we have no evidence relating to the entire class which it denotes.  But, if the conclusion distributes that term, it asserts something about the whole class, and so goes beyond the evidence."[3]

A term is distributed when reference is made to the whole class of objects it signifies [all or none] otherwise it is undistributed:  "Only Universal Propositions distribute their subject, and only Negative Propositions distribute their predicate."  What is the Law of Contradiction?"  It states that the same predicate cannot be denied and affirmed of the same subject.  What is the Law of the Excluded Middle?  It states "a given predicate must either be affirmed or denied of a given subject--S must either be P or not be P.[4]

### Obverse, Converse, Contrapositive, and Inverse

The **obverse** of a given proposition has the same subject but has a contradictory predicate.  The **converse** of a state proposition has its subject for predicate and its predicate for subject.  Not all propositions have a converse.[5]  What is a **contrapositive**?  It is the proposition derived by reversing and contradicting the subject and predicate of a given proposition.[6]  "Only the two universals have an **inverse**" or opposite.[7]  Thus, the inverse is a critical part of logic today.

20

### Logic, Dialectic and Physics

Two propositions are contrary if they affirm contrary predicates of a same subject, for example, good and bad. All syllogisms have a major and a minor premise. The major premise has the "middle term and the major term." The minor premise has "the middle term and the minor term." There are sixteen possible combinations of premises. The type of reasoning dealt with so far is qualitative deduction because it deals with the application of general propositions to particular cases.[8]

### Quantities and Deduction

Another form of deduction is quantitative deduction, for example, the Law of Gravitation which relates the force of attraction on two masses at a distance with a gravitational constant. Chains of syllogisms with a variety of degrees of complexity such as linear and systematic inference can be generated. Besides drawing mediate inferences from two hypothetical premises, mediate inference can be drawn from a "hypothetical major premise and a categorical minor premise."[9]

### Induction

Induction is the reverse of deduction. It is the inverse of deduction. A proposition infers from scientific facts. Science begins with classification and description. Classification needs description. For example, there is a variety of clover which has flowers with some different florets. A standard deviation of a sample of clover flowers with florets can be used to average the squares of the deviations from the arithmetical mean of a group. Evolutionary methods which show the stages where a development has occurred. An example is the method of drawing inductive inferences about a process. A second method, the comparative method, does not trace the evolutionary development of a phenomena, but, compares phenomena, such as comparative sociology. A scientific hypothesis is a definite supposition about phenomena that can be affirmed or denied by observations.[10]

### Several Inductive Methods

### Difference

If two sets of data are alike except that one is positive and another negative, then the result will follow this condition. For example, one piece of litmus paper turns red in acid and another dipped in water does not, then the turning red is the result of the acid.[11]

Logic, Dialectic and Physics

## Agreement

If several instances of an occurrence have a common cause, then that mutual antecedent "is a condition of that phenomenon." For example, if dew occurs on surfaces that are similar in all aspects except that they have a lower temperature than the surrounding air, then the lower temperature created the dew.[12]

## Residues

The weight of coal in a truck is determined if one knows the weight of the empty truck and deducts it from the weight of the loaded truck. The weight of the coal is the residue.[13]

## Statistics

Descriptive and correlation statistics can show patterns of association.[14]

## Deductive-Inductive Methods

Newton hypothesized that terrestrial gravitation determined the moon's orbit. Inductive methods could not determine this alone. Using deductive reasoning and calculus, Newton concluded that the moon should be deflected from its path by almost 16 feet per minute. This result Newton tested and determined the moon's orbit and period.[15]

## Deductive and Inductive Probability

Calculating deductive probability has certain stipulations: (1) we must know the number of exclusive alternatives. (2) These alternatives must have an equal chance of occurrences. (3) We must know how the alternatives favor the event. For example, the likelihood of throwing a head with a coin is 1/2. Inductive probability bases itself on "chance" or odds. That is, if the odds are 11 to 25 then the probability is 11/36 of an event occurring.[16]

What does Hegel's dialectical approach suggest about formal and inductive logic? It suggests that these two kinds of logic can be synthesized,for example, in an inverse and synthesis. Bertrand Russell and Alfred North Whitehead tried to synthesize logic and mathematics. They made some errors, but, it is clear there is truth to absolute numbers and absolute proofs in mathematical logic. There may be more than one absolute in

**Logic, Dialectic and Physics**

mathematics and physics, but, there is no getting around the existence of absolutes, for example, "Reproduction is the beginning of death."[17] God is beyond subtlety. Maybe formal and inductive logic are a unity.

Logic, Dialectic and Physics

## The History of Logic:  Logic as Complexity

Abelard (1079-1142):  The stoics called logic dialectics and this term was preserved in history as Dialectical or the logic of Abelard.[1]

Aristotle (384-322 B.C.):  In his "Organon" he developed two avenues of analysis:  Analytica posteriora and Analytica priora.  In Analytica posteriora Aristotle developed a series of propositions or true statements that fell into two classes. The first class is axioms or incontestably true statements. The second class is theorems whose truth can be shown by their relationship to the true axioms.  The subordination of these statements to the classes of correct statements obtains by working rules which we now call logic.  He formalized these rules in Analytica priora.  An example is "All S are P."[2]

Bolzano (1781-1848):  He built logic to an apex that would lead to symbolic logic.[3]

Brentano (1838-1917):  He reinterpreted Aristotle's elementary forms and reformed his syllogism.[4]

Frege (1848-1925):  He interpreted mathematical ideas in  concepts of logic, and he formalized logical calculus.[5]

Hegel (1770-1831):  His ideas on logic are presented in this text.

Hume (1711-1776):  He developed German "epistemology" and the "theory of science."[6]

Husserl (1859-1938): He laid the base for the theory behind logic.[7]

Kant (1724-1804):  He invented transcendental logic where he maintained that if "we presuppose that there is one an only one object have the particular properties of the thing described, this cannot obviously not be maintained in the case of logic."  Also, there are different types of logic.[8]

Leibniz (1646-1716):  He and Isaac Newton invented calculus.[9]

Locke (1632-1704):  He and Berkeley and Hume developed the "theory of science."  They laid the base for the field of the philosophy of science.[10]

Mach (1838-1916):  He fought metaphysics with science.[11]

### Logic, Dialectic and Physics

Russell (1872-1970):   He tried in his Principia Mathematica to convert mathematics to logic through inferential rules.  Russell developed "the first perfect formal logic.[12]

Schlick M. (1882-1936):  He helped invent logical positivism within the Vienna Circle during the early part of the twentieth century.[13]

Whitehead (1861-1947):  He helped Bertrand Russell in writing the Principia Mathematica which Russell published during the years 1910-1913.[14]

Wittgenstein (1889-1951):  He developed tautologies and "perfect forms" in logic where "we may assert universal validity for them."  In his Tractatus Logico-Philosophicus, he criticized Russell's work.[15]

There are many facets to logic, and the one emphasized here is dialectics.  Symbolic logic or the syllogism can also explain physics.  Symbolic logic applies to dialectics.  Dialectics apply to symbolic logic.

Logic, Dialectic and Physics

## A Brief History of Physics

Keep in mind the idea of dialectic while reading this essay on the history of physics. There is a dialectic and a new synthesis from Democritus to Rutherford, from Aristotle to Galileo, from Copernicus to Newton, from Newtonian physics to quantum mechanics, from Newton to Einstein, from amber to the monopole, from small particle physics to unified field theory. Many new research advances in physics came from small discrepancies in research like the photograph of the position track. A logic of the dialectic is a key to the evolution of the field of physics.

Greeks who pursued knowledge through reason were the <u>philosophers</u> or wisdom lovers. Philosophers who studied nature instead of human behavior, ethics, morality and motivation were <u>natural philosophers</u>. Today people term this field <u>science</u> from the Latin word "to know." The word physics is a shortened form of physical philosophy; it initially included all science. Physics includes such things as "motion, light, sound, electricity and magnetism."[1] Today physics has a wide intellectual agenda. It is a model discipline.

### Physics in Antiquity

<u>Anaximander</u>: He introduced the idea of 'first principle,'and, the idea that reality is one thing. It is unlimited and it is in motion.[2]

<u>Heraclitus</u>: He contended that we exist and we do not exist. Harmony consists of opposite tensions like a bow and arrow.[3]

<u>Pythagoras</u>: He coined the term 'philosophy,' and, he suggested that the first principles are numbers that contain harmonies called geometry.[4]

<u>Zeno</u>: He opposed the idea of motion saying, 'A moving body does not move in the place in which it is, nor in that which it is not.'[5]

<u>Anaxagoras</u>: He contended that 'Minds' govern the "motion of matter."[6]

### Archimedes and the Lever

In his book, <u>On the Equilibrium of Planes</u>, Archimedes creates the law of the lever and the problem of finding the center of gravity of an object. Then, he applies the principle of the lever to the pulley to move a larger ship. Following this, Archimedes invented a "law of floating bodies." It states "any solid body submerged in a liquid loses the weight of this liquid displaced by it."[7]

Logic, Dialectic and Physics

## Alexandria

After Athens declined, Greek culture moved to Alexandria, Egypt. There, Euclid published his Elements of Geometry. In Alexandria, another scientist, Hero, invented the siphon and the steam jet engine. Also, Ptolemy analyzed the principles of the refraction of light.[8]

### The Greek Idea of Motion

Aristotle (384-322 B.C.) thought that the force put on a stone sent itself to the air and then the air carried the stone. Motion away from a natural place led to motion toward the natural place and natural motion brought the stone object to its natural place. Rest, Aristotle thought, was the natural state. Movement of heavenly bodies followed different laws than the movement of earthly objects.[9] Galileo proved an anti-Aristotelian theory of gravity. It was a dialectical discovery.

### The Classical Era: Plato and Aristotle

Plato: He insisted there is a rational explanation of the cosmos. Plato enunciated the idea of geometrical atomism: "All body has depth." He felt that both reason and observation are important in science.[10]

Aristotle: Sambursky suggests that Aristotle maintained there are "principles for the cognition of nature," that objects "fulfill a purpose." There is an infinite, that time affects objects, that objects "in motion must be moved by something." There is a prime mover that is not divisible, has no parts and no magnitude, that movement is largely circular. There are five elements, and there are fifty-five spheres that move and counteract movement in the planets.[11]

### The Middle Ages in Physical Thought

Rhazes: He argued there are four elements: earth, wind, water and fire and a void or vacuum. He also suggested there is absolute and relative time and place.[12]

Avicenna: He maintained "every motion occurs through a power in the moving object by which it moves. "This power is either violent, or accidental, or natural." This opposes Aristotle's idea of "forced motion."[13]

Averroes: He criticized Ptolemy's system of spheres.[14]

## Logic, Dialectic and Physics

Moses Maimonides: He argued for Ptolemaic astronomy as opposed to Aristotelian physics. Thus, everything does not revolve around the earth.[15]

Roger Bacon: He argued that mathematics is the foundation of science. Bacon also suggested that vehicles might move through mechanics.[16]

Galileo Galilei (1564-1642) using inclined planes and free fall experiments found that heavy bodies did not as Aristotle thought fall more rapidly than light bodies. Galileo found that all bodies, regardless of weight rolled down planes or fell freely at "equal and constant acceleration."[17]

### "Let There Be Newton"

During the year that Galileo died Isaac Newton, a premature baby, was born to a Lincolnshire farmer. He was a backward student, but, after a fight with a schoolmate he went on to be first in his class. In 1665 Newton received a B.A. in mathematics from Trinity College. During 1665 a plague struck London killing nearly ten per cent of the population. The University of Cambridge closed because of the plague and Newton went home to Lincolnshire. There he invented a binomial theorem and the fields of differential and integral calculus. Also, Newton developed a theory of color and the law of gravity. At the age of 26 he became a professor at Cambridge. At the age of 30 Newton became a Fellow of the Royal Society. He published his work on mechanics and gravity when Newton was 44, and he published his work on optics at age 65. His Mathematical Principles of Natural Philosophy had its preface dated May 8, 1686.[18]

Isaac Newton (1642-1727) systematized Galileo's falling body experiments. Newton formulated three Laws of Motion in his Principia. What are these laws? The first law of motion is as follows: "a body remains at rest or, if already in motion, remains in uniform motion with constant speed in a straight line, unless it acted upon by an unbalanced external force." The second law of motion is thus: "The acceleration produced by a particular force acting on a body is directly proportional to the magnitude of the force and inversely proportional to the mass of the body." The third law of motion reads as follows: "Whenever one body exerts a force on a second body, the second body exerts a force on the first body. These forces are equal in magnitude and opposite in direction."[19]

What did Newton discover on gravity? He did not just find that all objects on earth attract it. Newton argued "all masses attracted all other masses." Thus, the gravitational attraction of the earth is not unique. There is a gravitational force between any two physical objects in the universe. Furthermore, as the distance between two bodies increases the force of gravity between them is inversely proportional to the square of this distance. Scientists discovered the gravitational constant needed in the equation.[20] Today scientists have not found a gravity particle.

## The Conservation of Momentum

Is there a law about momentum? Yes, "the total momentum of an isolated system of bodies remains constant." Moreover, "the total angular momentum of an isolated system of bodies remains constant." Momentum is conserved.[21] Are there systems where momentum is not conserved? Perhaps.

## From Copernicus to LaPlace

Copernicus: He argued that the heavens were immense in relation to the size of the earth. Copernicus developed the idea of a heliocentric system.[22]

Giordano Bruno: Bruno argued for "an infinity of worlds." He maintained that the universe did not have a center or a circumference and that are an "infinite number of solar systems in the universe."[23] There probably is more than one solar system.

Johann Kepler: He suggested that the structure of the universe is harmonic.[24] Kepler, in the sixteenth century, maintained "all planets follow elliptical orbits with the sun located in one of the foci."[25]

Galileo: He stated that scripture and nature are "two aspects of one truth." Also, sun spots exist. Galileo developed the idea of inertia. He criticized Aristotle's idea of 'natural motion.' Galileo used the telescope to back Copernicus' theory. He tried to measure the velocity of light. His main contribution was the idea that all bodies fall with the same velocity. This idea refuted Aristotle.[26] Galileo discovered the law of the pendulum. It states that the duration of the swings of a pendulum remained the same even as their duration becomes shorter. He also found that light and heavy bodies fell at the same rate.[27]

Descartes: He suggested that the nature of a body consists in extension. There is nothing divisible in thought, "which we do not recognize to be divisible." Thus atoms do not exist.[28]

Logic, Dialectic and Physics

Pascal:  Air has weight as a whole mass and it presses on surrounding bodies.[29]

Leibniz:  He argued that space and time are relative.[30]

Euler:  He asserted there is positive and negative electricity.[31]

LaPlace:  He helped found probability theory.[32]

## The Law of Energy

What is the law of energy?  James P. Joule (1818-1819) found that a fixed amount of one type of energy changes into a fixed amount of a different type of energy.  Thus, energy is neither created or destroyed.  Hermann von Helmholtz (1821-1894) formalized this law.  Later in 1905, Albert Einstein proved that mass is a form of energy.  In 1931 Wolfgang Pauli (100-1958) developed the idea that a subatomic particle, called the neutrino, accounted for discrepancies from this law.[33]  Energy is a form of mass.

## Vibratory Motion

The law of conservation of energy holds for vibratory motion, for example, harmonic motion like a branch in the wind, and simple harmonic motion, that is, a vibrating violin string.  Pythagoras of Samos (sixth century) studied the relationship of vibrations to music.  Today we call these vibrations harmonic motion.  As was mentioned earlier, Galileo found that the period of a pendulum's swing did not result from the weight of its bobs but upon the square root of the string's length.  Recently, atoms' vibrations within molecules using the rules found in simple harmonic motion measure time with great accuracy.[34]  Kepler thought that the spheres of the universe created music-- "the music of the spheres."  Scientists are trying to relate vibrations of small particles to the motion of large bodies of matter and energy.

## Of Liquids

Liquids take on the shape of their containers under the downward gravitational force.  Gases do not have a definite shape or volume.  Pressure is the force per unit volume.  Density is the weight for each unit volume.  A fluid puts pressure in all directions.  Can we make a rule on pressure?  Yes, we can.  It is Pascal's principle:  "Pressure exerted in any place on a confined liquid is transmitted to all portions of the interior unchanged."  Hydrodynamics is the mechanics of liquids.  Pneumatics is the mechanics of gases.  Together they make up <u>fluid mechanics</u>. Buoyancy is the upward force liquids exert against submerged objects. <u>Archimedes principle</u> states this process. Surface energy is the potential energy of the liquid pushed upward from its surface. "Hydrodynamics" is the study of how the pressure of a fluid falls as its velocity grows larger.[35] Are there situations where the pressure of a liquid does not fall as its velocity increases? In theory this anti-system is possible. Blaise Pascal and Daniel Bernoulli invented the basis of hydrostatics. It is a law which states that a fluid in a closed container presses with equal force on each area at any part of the container. Thus, the force of a hand to a piston in a small cylinder creates a larger force on a wider cylinder. It can raise a carriage on the larger piston.[36] What is Bernoulli's principle? It is "the statement that an increase in the speed of a fluid produces a decrease in the pressure and a decrease in the speed produces an increase in pressure."[37] For example, the air moving over an airplane wing has higher velocity and lower pressure than the air moving under the wing. The difference between these pressures creates the lift of the airplane.[38]

## Gases

Democritus thought that matter cannot divide completely. The smallest portion is the atom. He invented <u>atomic theory</u>. Today there is an anti-Democritus synthesis in nuclear physics. At absolute zero, motion in gases comes to a standstill. Compared to liquids, gases are rarefied. Gases have differences in densities. In 1658 Pascal measured atmospheric pressure. Evangelisto Torricelli (1608-1647) created the barometer and the vacuum around 1644.[39] Today, linear accelerators are vacuums.

## Boyle's Law of Gases

If atomism is correct, particles of gas should be compressible. Robert Boyle (1627-1691) in 1660 found that for a specific quantity of gas, pressure relates inversely to volume. Matter and gases are today compressible. The product is a constant: $PV = K$. John Dalton, an Englishman (1776-1844) verified atomic theory by studying Boyle's experiments.[40]

## Sounds

Fluids, like water waves, can move. A wave is a moving distortion. Waves have a <u>crest</u> and a <u>trough</u>. The quantity of crests passing a point in a second is the "frequency of the wave."[41] Today an ear drum can help locate the position and frequency of very small particles Mike Franey suggests.[42]

## Logic, Dialectic and Physics

### Properties of Sound Waves

In the 6th. century B.C., Pythagoras of Samos studied sounds of plucked strings. He found that short strings vibrated faster than long strings. About 400 B.C. Archytas of Tarentum (420-460) suggested that the striking together of articles created sound. Quick motion created a high pitch and a slow motion develops a low pitch. Aristotle argued that sound could not transmit itself in a vacuum. In the 1st. century B.C. Marcus Vitruvius Pollio contended that near a vibrating string the air vibrated. These vibrations we listen to as sound. About 500 A.D., a Roman, Anicius Boethius (around 480-524) compared sound and water waves.[43] Radio waves hit the earth from far distances. Thus, there is some truth to the idea of the music of the spheres.

How are water and sound waves different? Water waves are transverse waves. Sound waves carried through the air are not transmitted but are made "of periodic compressions and rarefactions." Compression waves move parallel to the square of propagation. The loudness of a sound decreases as it moves farther from a source. Loudness is measure of the amount of energy that passes through one square centimeter of area. The area is perpendicular to the direction of the sound. The more air compressed the more energy and the louder the sound. The intensity of sound varies inversely with the square of the distance from its source.[44]

Pitch is shrillness in tone. As frequency goes up, the sound becomes more shrill. The sound deepens as frequency decreases.[45]

The ratio of the velocity of an object to the velocity of sound in a different medium is the Mach number named after Ernst Mach (1838-1916), an Austrian. To fly at "Mach 1" means to equal the velocity of sound, 758 miles/hr.[46]

Ernst Mach: He criticized "Newton's absolute space."[47]

Different pitches produce various sounds on a musical scale. Pythagoras found that notes fit together that have tiny whole-number ratios. Pitches change if their source moves in relation to the listener's source. The Doppler effect is the name of this effect after Christian Doppler (1803-1853), an Austrian, who explained it in 1842. Timbre is the difference in notes of same pitch. Also, sound waves reflect and in some cases bent around obstacles that are of its wavelengths. Augustin Fresnel (1788-1827), a Frenchman, discovered this phenomenon.[48] The keys on a piano form a Fibonnaci series, for example, the addition of black and white keys equals thirteen, which forms a Fibonnaci number. This is a proof for Platonic idealism.

## Physics Thought from Dalton to Clausius

Dalton:  His main contribution was the evaluation of the relative weights of "ultimate particles."[49]

Augustin Fresnel:  He looked at the wave theory of light and studied interference, diffraction and polarization.[50]

Hans C. Oersted:  He discovered that electricity moved a magnetic needle.[51]

Andre Ampere:  He showed how two electric currents interacted.[52]

Sadi Carnot:  He tested how heat produces motion.[53]

James Joule:  He determined the mechanical equivalent of heat.[54]

Hermann Helmholtz:  He studied how force conserves.[55]

Rudolf Clausius:  He analyzed heat transfer and argued "heat cannot by itself pass from a colder to a warmer body."[56]

## Hot and Cold Temperatures

The temperature of an object is its degree of coldness or hotness.  Many materials change in volume as the temperature changes.  For example, glass expands or contracts as temperature changes.  Bimetals and mercury can measure temperature changes.  During 1714 Gabriel Fahrenheit (1686-1736) invented the mercury thermometer.  Guillaume Amontons (1663-1705) studied the expansion of gases as temperatures changed.  He found that if one encloses a gas as the temperature increases the pressure also increases.[57]

During 1802 Joseph Gay-Lussac (1778-1850) found that air, oxygen, nitrogen and hydrogen have similar coefficients of cubical expansion.  Absolute zero or 273 degrees Celsius is the lowest temperature.[58]  At this temperature motion virtually stops.  The temperature of the universe is only three degrees Kelvin which is extremely cold.  Today so-called superconductors operate near absolute zero.

## Logic, Dialectic and Physics

What is Gay-Lussac's law? It is "the volume of a given mass of gas is directly proportional to its absolute temperature, provided the pressure on the gas is held constant." How are volume, temperature and pressure related? For any quantity of gas the "volume times the pressure divided by the absolute temperature remains constant: $PV = RT$."[59]

### "Heat"

The kinetic theory of gases is the idea that gases contain particles in motion. Bernoulli contended in 1738 that gas particles bouncing off a container's wall create pressure. During the 1860's James Clerk Maxwell (1831-1879) a Scottish scientist and Ludwig Boltzman (1844-1906) proved the kinetic theory of gases. Maxwell wrote an equation that tested the distribution of gas particles at different temperatures. There are a variety of kinetic energies. If we discern the average velocity of the gas, we can get the average kinetic energy. The particles have both low and high energies.[60]

What is heat? It is the internal energy related to the random motion of atoms that form matter. Thus absolute temperature measures the average kinetic energy of the individual atoms or particles of a system. At absolute zero a particle's kinetic energy is very small. Diffusion is the ability of two gases to mix. Thomas Graham (1805-1869) found in 1829 that the rate of a gases' diffusion is inversely proportional to the square root of its density. It is a function of its molecular weight.[61]

What is the specific heat of an element like iron? It is the quantity of heat needed to increase the temperature of one gram of an element like iron 1 Celsius degree. The heat consumed in melting is latent heat.[62]

### Heat and Energy

Galileo invented the thermometer in 1592.[63] James Black (1728-1799) saw heat as a fluid or "color" which can penetrate material bodies raising their temperature. Sadi Carnot, also, had this idea.[64]

### Heat as Motion

Benjamin Thompson, an American, born during the Revolutionary period had the concept that heat was the internal motion of a material.[65] Count Rumford, Julius Mayer and James Joule depicted the mechanical heat equivalent during the first part of the nineteenth-century.[66] The first law of thermodynamics equates heat and mechanical energy. Rudolf Clausius and

34

Logic, Dialectic and Physics

Lord Kelvin formed laws explaining how one form of energy transforms into another.[67]

Ludwig Boltzmann, James C. Maxwell and Josiah Gibbs developed the idea "that heat is the energy of motion of tiny particles." Today Brownian motion or the motion of plant spores on water was given a mathematical expression by Albert Einstein. Brownian motion proved the kinetic theory of heat. Einstein described the statistical theory of heat.[68]

Hot bodies emit light: "as the temperature goes up, the emitted radiation becomes rapidly more intensive, and richer in the short wave lengths."[69] Two laws relating about light propagation by hot bodies appeared during the second half of the nineteenth century. Wien's law states "that the wave length corresponding to the maximum intensity in the spectrum is inversely proportional to the (absolute) temperature of the emitting hot body." The Stefan-Boltzmann law says "the total amount of energy emitted by a hot body is proportional to the fourth power of its (absolute) temperature."[70]

Hot gases send forth light of specific wavelengths. In contrast, to solids or liquids which emit a continuous spectrum of wavelengths.[71] Gustav Kirchoff (1787-1826) found "all substances absorb the same light frequencies which they can emit. Joseph Fraunhofer (1787-1826) using a prism found thin black lines coming out of a rainbow experiment. These "Fraunhofer Lines" arise when the continuous spectrum of the photosphere or dense part of the sun passes through the chromosphere where wave lengths of the chemical elements absorb and scatter and create the dark lines.[72]

## Heat in Motion:  Thermodynamics

Previously mentioned, the first law of thermodynamics says that the total energy in a closed system is constant. The second law of thermodynamics states that in a closed system heat flows from the hot area to the cooler area. Rudolf Clausius (1822-1888) discovered this law in 1850. It applies to all forms of energy. Clausius argued that entropy is a measure "of the unavailability of energy." It is "heat divided by temperature." By these laws the total energy of the universe remains constant and entropy or disorder is increasing. Energy in the form of motion, sound and heat relate. "Sound and heat are forms of kinetic energy."[73] Is the universe becoming more and more disordered? John O'Dougherty thinks "the mind seeks order."[74] Mike Franey thinks that computer chips are the product of a probe for order. These chips organize data more and more.[75] This is a new principle of order in the universe.

Logic, Dialectic and Physics

## The Transmission of Light

After Newton, physicists varied between the wave and the particle theory of light. In the twentieth century these two views became one. A dialectic resolved. Light can cross a vacuum. Light moves in straight lines. A light source radiates light. It can reflect from opaque bodies. Refraction is the breaking back of light moving from one transparent medium to a different medium. For example, the index of refraction of air is minuscule.[76]

## Of Lenses

In a prism light bends first on entering the glass and then on leaving the glass. Convex lenses converge light. Concave lenses diverge light. What is the lens formula? It is $1/Di = 1/f - 1/Do$ where Do is the light source and Di is the image and f is the focus. In the eye the "cornea and crystalline lenses converge the light rays to focus upon the light sensitive inner coating (retina) of the rear of the eyeball."[77]

Roger Bacon (1214?-1294) was one of the first people to use eyeglasses. They came into existence during in the Middle Ages. Benjamin Franklin (1706-1790) invented bifocals. During 1599 Giambattista della Porto (1538?-1615) created the camera. In 1889 Thomas Edison (1847-1931) invented the motion picture camera. Anton Van Leeuwenhoek (1632-1723) and Zacharias Janssen created the compound microscope which had several lenses. An apprentice of a Dutch spectacle maker Hans Lippershey invented the telescope around 1608. Galileo Galilee (1564-1642) used a telescope to see mountains and craters on the moon, Venus and Jupiter, and the satellites of Jupiter. Scientists built in 1897 a telescope with a 40-inch lens at the Yerkes Observatory in Wisconsin.[78]

## Of Optics

Mentioned before, Newton's, Optics, describes the different refractivities of the light of different colors. He maintained that white light contains rays of different colors and refractivities. Newton invented the "spectroscope." The rainbow is an example of refraction of white light by water droplets. The color arches around the moon are the result of reflection "(not refraction)" of ice crystals in cirrus clouds. Newton built a reflecting telescope in 1672. He created a corpuscular theory of light in contrast to Huygens' wave theory of light. Thomas Young and Augustin Fresnel developed experiments supporting Huygens' wave theory of light. During 1705 Newton became a knight. He

36

died at age 85 in 1727. Newton was academically unproductive for the last 25 years of his life.[79]

## Of Color

Light has color, for example, the rainbow. Lucius Seneca (4 B.C.?-65 A.D.) argued that the rainbow was like the colors seen at the side of a piece of glass. Rene Descartes defined the rainbow mathematically. In 1666 Isaac Newton refracted light in a prism. He argued that light was a mixture of colors. During 1807 Thomas Young (1773-1829) and Herman von Helmholtz (1821-1894) suggested that red, green and blue could if properly combined create the sensation of all other colors. During the years 1814 and 1824, Joseph von Fraunhofer (1787-1826) working with prisms found dark spectral lines. He placed a prism at the eyepiece of his telescope and in so doing invented the spectroscope. After this, Gustav Kirchoff (1824-1887) combined the spectroscope with chemistry and began the science of spectroscopy. He passed the lights of different elements through his spectroscope and discovered emission spectrum or a few images of the light in the slit. These lines fingerprint the elements.[80]

Robert Bunsen (1811-1899) discovered two new elements through spectroscopy. Kirchoff studied glowing solids which emitted all colors of light and formed a <u>continuous spectrum</u>. For example, if the light of a carbon-arc standing for a continuous spectrum passes through a sodium vapor at a cooler temperature than the arc, the sodium vapor absorbs some of the light. The light absorbed is of the variety of the light that sodium vapor would emit if it glowed. The sodium vapor creates two yellow lines that composed its spectrum. When this cool sodium absorbs the light coming out of a continuous spectrum two dark lines cross the spectrum in the area of bright lines (two) of the sodium emission spectrum. They are the sodium <u>absorption spectrum</u>. The sun is an absorption spectrum. Scientists found, in this manner, helium's existence in the sun.[81]

What is diffraction? Diffraction is the bending sideways at the ends of a wave front. Francesco Grimaldi (1618?-1663) saw the diffraction of light during 1665. He sent light through two openings and showed that the light was wider than it would have been if it had traveled through two openings in an straight line. Thus there was diffraction. Also, there were color effects, red light, for example, had the most diffraction. These findings fit the wave theory of light which Christiaan Huygens (16629-1695) championed. Newton formulated the particle theory of light. Diffraction gratings formed spectra.[82]

Logic, Dialectic and Physics

## Of Light Waves

During 1801 Thomas Young showed that light developed interference patterns like those found in water waves. He calculated the wavelengths of light from these patterns. During 1923 Albert Michelson (1852-1931) calculated the speed of light to within nearly ten miles per second of the presently accepted value of 186,281.7 miles per second.[83] It is a universal constant argued Einstein.

During 1842, Christian Doppler pointed out that in an approaching light source light waves crowd together and are of a higher frequency. In a receding light source the waves of light pull apart and thus are lower in frequency and redder. In 1800 a British astronomer, William Herschel (1738-1822) located infrared radiation. In 1801 Johann Ritter (1776-1810) discovered ultraviolet radiation. Later, in 1808 Etienne Malus (1775-1812) found polarized light.[84]

## Of Ether

Aristotle thought that a rarefied gas called ether made up heavenly bodies. Scientists thought that it might transmit gravity or light. Michelson-Morley qualified this theory in 1887. Relativity stopped the need for a theory like ether. Einstein argued that gravity is a property of space-time geometry and not a transmitted force.[85]

## Relativity

First, Einstein suggested that all motion is relative to some object at rest; and, any object can be taken with equal truth as the frame of reference. Thus, no object is "really" more at rest than any other object. All motion is relative. Second, he assumed that the velocity of light in a vacuum would be constant and besides the motion of the light source to the observer. The velocities of an object will not exceed the velocity of light.[86]

In 1905 Einstein equated mass and energy. This was a novel approach because earlier chemists with the law of conservation of mass that mass could not be created or destroyed. Einstein argued for an interchange of mass and energy or $e = mc2$ where $c2$ is the velocity of light. Also, he maintained that time proceeds more slowly as motion increases. Scientists call this phenomenon time dilation.[87]

## A Dialectical or Paradigm Shift: Relativity

## Logic, Dialectic and Physics

Einstein tried "to geometrize all physics."[88] He thought if there is not ether filling the universe, then absolute motion does not exist because one cannot move in relation to nothing. Motion is relative. The speed of light is constant independent of the speed of its source.[89]

Newton maintained that space and time are independent entities. Einstein unified space and time. He predicted time dilation or clock slowing when seen from a system that is moving.[90] Also, an electromagnetic field is a physical entity in an empty space.[91] Light has pressure.[92] Time is a fourth dimension.[93]

### History

Albert Einstein was born in Ulm, Germany by Munich on March 14, 1879. His father owned an electrotechnical business in Ulm. He lived in Munich during his youth and then he went to the Polytechnical School in Zurich, Switzerland. In 1901 he became a patent examiner in Bern. During 1905, when he was 26, he published three articles on three areas of physics: "heat, electricity, and light" in the German magazine, Annalen der Physik.[94] The rest is history.

### Einstein's General Theory

First, Newton's second law of motion used the idea that mass defined the quantity of inertia held by a body. Second, Newton also defined mass as the power of a gravitational field an object generates. Einstein equated inertial and gravitational mass. Thus, the mass of a particular body interacts with the mass of all other masses in the universe. He argued that mass increases with motion.[95]

### Gravitation

Einstein argued that matter would curve space and that gravity would affect light rays.[96] Today there are many theories on gravity. The writer thought up independently the idea of anti-gravity.

### Of Quanta

A so-called black body would be a chemical substance that would absorb light of all frequencies. It would not reflect light. During 1879 Josef Stefan (1835-1853) found that the energy sent forth by a body increases as a fraction of the absolute temperature. During 1884 Boltzmann gave a mathematical proof that this was true for black bodies. In 1895 Wilhelm Wien (1864-1928), because there is no ideal black body existing, used a furnace with a hole in it to stand for a black body. Radiation coming out of the hole would be black-body radiation. Light would initially go into the hole. Then, the furnace would absorb it. Wien found that as he raised the temperature the energy radiated had higher frequencies. The top frequency varied directly "with the absolute temperature." Thus, the colors of stars are a function of their temperatures.[97]

There is a problem, however. Why did the probability of radiation decrease as frequency increased? Max Planck (1858-1947) argued that energy did not move continuously but moved in discrete packets of energy called quanta. He found that the energy of a quantum of radiation is "proportional to the frequency of that radiation," or $e = hv$ where h is Planck's constant. Ultraviolet light (a light of high frequency) could produce chemical reactions more easily than the small quanta of a lower frequency red light.[98]

## Of the Photoelectric Effect

What is the photoelectric effect? It is the ejection of electrons from metal surfaces under the impact of light. In 1905 Einstein explained this phenomena by arguing that light radiated and absorbed in quanta. He developed the following equation: $1/2\ mv2 = hv\text{-}w$. $1/2\ mv2$ is the kinetic energy of the electron emitted; "hv (Planck's constant times frequency) the energy content of the quanta being absorbed by the surface; and w the energy required to break the electron." Furthermore, Einstein argued that light consists of particles of energy or photons. "A photon is both a particle and a wave."[99]

## Of Magnetism

Thales (640?-546 B.C.) was one of the first Greeks to study magnetism. Later, he found that amber after rubbing created an attractive force. Thales studied iron-attractive materials and called them ("the Magnesian rock"). He also studied amber which had a force that differed from magnetism because it could attract light objects like feathers or leaves. Eventually, the word electricity came to identify this phenomena.[100]

Later, people found that magnetism transfers. For example, steel stroked with magnetic iron can become a magnet. Also, the ends of a magnetized needle point towards the poles of the earth. The first applications of a magnetized needle, a compass, used in ocean voyages came during the twelfth century. People found Like poles repel; unlike poles attract."[101] Some modern day physicists think that a monopole exists.

## Logic, Dialectic and Physics

During 1785 Charles Coulomb (1736-1806) discovered that magnetic force relates **inversely** to the square of the distance from the object. This is similar to gravitational attraction. After WWII, ferrites came into existence. They are a mixture of oxides of iron and other metals like cobalt and magnesium. These ferrites can produce ferromagnetism. Magnetic domains are the areas that have magnetic forces concentrated. For these ferromagnetic substances the temperature which disrupts magnetic domains is the Currie point after Pierre Currie (1859-1906).[102]

The earth is a magnet. It is not known why it is a magnet. The earth has North and South magnetic poles. Michael Faraday (1791-1867) discovered magnetic lines of force. A magnetic field is the area that a magnetic field changes the geometry of the surrounding space. The strength of the magnetic field is the magnetic flux density. During 1845 Faraday found that magnetic poles repel rubber, glass and sulfur. Today scientists call them diamagnetic substances.[103]

### Of Electrostatics

William Gilbert (1540-1603) invented the earth magnet idea. Furthermore, he analyzed the forces created by rubbing amber. Gilbert called the substances that created attractive forces after rubbing, "electrics." Later, people thought that electricity was a fluid. The charge gained from rubbing appeared to be stationary or static electricity. Electrostatics was the field that studied these properties. Otto Guericke (1602-1686) found similarities between magnetic forces and static electricity.[104] What might a dialectical approach suggests about positive and negative charges? Maybe a quasi-charge is possible!

Stephen Gray (1696-1736) and Charles DuFay (1698-1739) studied electrostatic attraction and repulsion. Benjamin Franklin developed the ideas of positively charged and negatively charged bodies. Today we have the law of conservation of electric charge: "electric charge can neither be created nor destroyed. The total net electric charge of the universe is constant." Scientists discovered subatomic particles and found that certain particles contained an electric charge.

The particles that have an electric charge are the proton and the electron. Coulomb's equation, mentioned earlier, expresses the relationship $F = qq/d2$. where q and q' are two bodies, F is the Force and d is the distance between the bodies.[105]

Logic, Dialectic and Physics

Electromotive force is the charge that moves across an electric potential difference. Piezoelectricity is the electricity created through pressure. The so-called Leyden jar is a condenser that can produce a long shock. Benjamin Franklin invented the lightning rod to bring lightening safely to the ground.[106]

## Of Electric Currents

Alessandro Volta (1745-1827) found that a spark drawn from the crown of a "voltaic pile" could flow to the top and a continuous current could flow through this wire. A "voltaic pile" is a pile of metal discs and cardboard discs wetted by salt water. Scientists call it the Volta "pile." Faraday labelled the metal contact points set in solutions, electrodes. Today scientists call the study of chemical reactions that create electric currents electrochemistry. Electrolysis is creating chemical reactions using electricity. Resistance is the ratio of potential difference to current intensity or $R = E/I$. **Semiconductors** are substances that have only moderate resistance. The equation for electric power is $P = EI$ where P is power, E is the potential difference, and I is the current intensity. What is an electric circuit? It is the pathway from one pole of a battery to another. The pathway after leaving a battery must return to another from a circuit.[107]

## Of Electromagnetism

Until early in the nineteenth century people thought that electricity and magnetism were different forces. In 1819 Hans Oersted (1777-1851) ended a lecture by placing an electrical wire that carried a current over a compass. The needle moved violently. The rise of the field of electromagnetism followed. Later, Dominique Arago showed that a flow of electricity was magnetic. In 1820 a Frenchman, Ampere, showed that two wires carrying an electric current attracted or repelled each other. The earth's center behaves like a solenoid as opposed to a bar magnet. Solenoid means "pipe-shaped." The more coils of a wire or solenoid has the more lines of force occur on the interior. The magnetic flux relates directly to the "number of coils (N) and inversely with the length (L)." It is proportional to N/L.[108] Also, different planets have different magnetic fields.

In 1823 William Sturgeon (1783-1850) wrapped copper wire around a bar that was U-shaped and created the electromagnet. During 1831 Joseph Henry (1797-1878) using insulated wire developed an electromagnet that could move a ton of iron. In 1844 Henry helped Samuel Morse (1791-1872) create the telegraph. In 1876 Alexander G. Bell(1847-1872) used electromagnets to create the telephone. Michael Faraday in 1831 used a magnet to induce a current, in so doing, he found <u>electromagnetic induction</u>. Shortly after this, he found that an electric current arose when a magnetic force moved across an electrical conductor. He devised a system where magnetic lines are continuously cut. In so doing, he generated current without the aid of chemicals and built the first <u>electric generator</u>.[109]

## Of Alternating Currents

Faraday's generator used a copper disk turning between a magnetic pole. Today a generator has copper wire wound on a spinning iron drum between electromagnetic poles. The turning coils are the armature. <u>Direct current</u> is when a circuit in a generator flows one way. In contrast George Westinghouse (1846-1914) backed the development of the alternating current.[110]

Edison and Kelvin approved the use of alternating circuits but alternating currents can transmit current over long distances better than direct current. During 1831 Faraday invented the transformer. How does a transformer work? If the current in a primary coil maintains a potential difference of 120 volts and if the secondary coil has ten times as many wire turns as the primary coil, then a potential difference of 1200 volts will occur for induced current. Following these inventions alternating current went to every factory an home. Falling water spun turbines to turn armatures and create electricity. What is a motor? It motion created out of current. Motors used an independent wheel powered by a generator.[111]

## Of Electricity

George Ohm developed the idea of electric <u>resistance</u>.[112] Michael Faraday discovered chemical decomposition by electricity. He found that a magnet induces a current of electricity in a coil when he pushed the magnet in and then pulled it out. Faraday found that a magnetic field can affect light. He also developed the idea of magnetic poles.[113]

James Clerk Maxwell gave mathematical expression to Faraday's discoveries. He also proved that an oscillating magnetic field can send out waves of energy. Heinrich Hertz confirmed this phenomenon in 1888 and led to the development of the radio. Maxwell created the electromagnetic theory of light.[114]

## Of Electromagnetic Radiation

## Logic, Dialectic and Physics

During the 1860's Maxwell wrote four equations to describe the relationships of magnetism and electricity. He found it difficult to deal with a magnetic or an electric field alone--a single electromagnetic field existed. Maxwell contended that a varying electric field induced a changing magnetic field. He predicted electromagnetic waves with frequencies that varied as the electromagnetic field changed. Maxwell argued that light is electromagnetic radiation. Thus, electricity, magnetism and light were all part of the electromagnetic field. He predicted that electromagnetic radiation existed at frequencies far different from light. In 1888, Heinrich Hertz (1857-1894) discovered radio waves or electromagnetic radiation of low frequency. During 1895 Wilhelm Rontgen (1845-1923) discovered x-rays or high frequency electromagnetic radiation.[115]

## Electrons, Protons and Neutrons

### Atoms

Democritus was the first to develop the theory of atoms. Epicurus and Lucretius also expounded this theory. Robert Boyle (1627-1691) used experimentation and observation to the study atoms. Boyle suggested that one could change the volume of a gas without altering its mass. He argued that the universe contained simple substances called elements. Elements imply a truth for reductionism. A compound would be the union of elements. Lavoisier (1743-1794) suggested that in a closed system a chemical reaction would leave the total mass the same. John Dalton (1766-1844) an English chemist, invented modern atomic theory. He contended that elements had atoms with the same mass, that different elements had atoms of a different mass, and that compounds created by the union of atoms into molecules like water $H_2O$. A chemical formula describes these molecules.[116]

In 1869 Dmitri Mendeleev (1834-1907) listed the atomic elements by atomic weight. Elements with similar properties, he placed in the same column or row. He found that properties of elements had fixed periods. He established the periodic table.[117]

What is molecular weight? "It is the sum of the atomic weights of the atoms" that make up a molecule. In 1811 Amedeo Avogadro (1776-1856) stated that equal volumes of gases have equal numbers of molecules under circumstances of fixed temperature and pressure. Today we can photograph some atom patterns.[118]

### Of Ions and Radiation

The decomposition of some molecules after an electric current passes through an electrolyte and creates chemical changes called electrolysis. Electroplating is the clinging of a metal to an electrode. Humphrey Davy and Michael Faraday were two of the first scientists to study these processes. Faraday developed two laws of electrolysis: (1) "the mass of element formed by electrolysis is proportional to the quantity of electric current passing through an electrolyte; (2) "one faraday of electricity will form an equivalent weight of an element when passing through a compound of that element."[119]

Does electricity like matter contain indivisible units? If this is true, and there are positive and negative units then a hydrogen ion moves to the cathode "by a positive electrical unit." In contrast, chlorine atoms move to the positive electrode and move by a negative unit.[120] There is a thesis and an anti-thesis present here.

In 1887 Svante A. Arrhenius (1859-1927) found that electricity could dissociate sodium chloride into positive and negative ions. Moreover, George Stoney (1826-1927) called the smallest unit of electricity, the electron. In 1869 Johann Hittorf (1824-1914) and William Crookes showed that in a vacuum tube or Crookes tube electricity cast a shadow "against the luminescence on the glass." Eugen Goldstein (1850-1930) said it was radiation and called it cathode rays during 1876. The particles of these rays were the "atoms of electricity."[121]

### Of Radiation

Heinrich Hertz (1857-1894) formed long wavelength radiation called radio waves. In 1800 William Herschel (1738-1822) discovered infrared radiation." Later, scientists found the radiation next to infrared radiation on the spectrum, called microwaves. In 1801 Johann Ritter (1776-1810) found shorter wavelength radiation called ultraviolet radiation. On January 23, 1896, Rontgen presented in a lecture one of the first x-ray photographs of a hand.[122]

### Of the Electron

During 1897 Joseph Thomson (1856-1940) recapitulated Hertz's experiment using a better cathode-ray tube. He detected a change in the path of cathode rays. Thomson thought these cathode ray particles represented unity of electricity. He called these negatively charged particles the name given

45

them by Stoney the "electron."[123]  Today, while we can measure the charge on an electron we do not still, as Roger Hellebuyck argues, know what an electron is.[124]


Robert Millikan (1868-1953) tested the size of the electric charge of the electron in 1911.  A field called electronics arose that studied free electrons and developed means for connecting them, for example, diodes and triodes. On December 12, 1901, Guglielmo Marconi (1874-1937) used Hertz's radio wave producing methods to send a radio wave from England to Newfoundland.[125]

Heinrich Hertz:  He argued "light is an electrical phenomenon."[126]

Edward Appleton (1892-1965) discovered a region of charged particles called the ionosphere in 1924.  Scientists modulated radio waves to send sounds. These waves had amplitude modulation or AM.  Later, Edwin Armstrong (1890-1954) developed a device to modulate the frequency of a radio carrier wave or FM.  During 1938 Vladimir Zworykin (1889-) created the iconoscope, a sophisticated cathode-ray oscillograph which used an electron beam that varied by a pattern on a fluorescent screen, the television tube.[127]

Using electromagnetic waves Robert Watson-Watt found that the time between the sending and return of a pulse's echo after hitting a reflecting object could determine the distance of the object.  Radar or "radio detection and ranging" was the name given to this device.  In 1932 Karl Jansky (1905-1950) found microwaves coming from the constellation Sagittarius.  After this, radio telescopes and radio astronomy came into being.[128]

### More on Atoms and Electrons

Previously, we discussed the photoelectric effect where light produced electricity from metal.  Were these electrons that came from the metal hit by light?  In 1899 Thomson studied these particles and found that they were identical with cathode ray particles.  Did these electrons always exist in the metal?  Did they exist between the atoms or within the atoms making atoms not the indivisible objects that Democritus and Dalton had formulated.  Is there relevance to an anti-Democritus' conception?  In 1898 Thomson contended that the atom was a solid sphere positively charged that had electrons stuck in it.[129]  Mike Franey suggests that this model of the atom was like a pie.[130]  In 1903 another scientist named Philipp Lenard suggested, in contrast, that the atom is mostly space.[131] Scientists displaced Democritus in a dialectic and new synthesis.

In 1906 Ernest Rutherford (1871-1937) hit a thin leaf made of gold with alpha particles. Some of the particles deflected. Rutherford showed there was an electric interaction. The alpha particles had a positive charge and the object within them also had a positive charge. Rutherford called the central object the atomic nucleus. Electrons composed the outside area of the atom. Ions would then have a gain or loss of electrons. His nuclear atom was almost empty. Rutherford suggested that the nucleus varied with the atom's atomic weight. Using x-rays Henry Moseley (1887-1914) found in 1913 that the simplest atom had a positive charge of 1 on its nucleus and the next a charge of two. He termed the size of the charge the atomic number.[132]

How are the electrons in an atom ordered? Scientists found that chemical reactions involved electrons held in shells within atoms. Electrons and electron shells explained the periodic table of elements.[133] Did Hegel predict the positron? Farber thinks that once scientists discovered the electron a positron could be found. The periodic table is logical. Chemical structure is often quite complex and subtle.

## Of Electrons and Quanta

### The Bohr Atom

In 1913 Niels Bohr (1885-1962) applied quantum theory to explain the problem that a circling electron not only gives off radiation, but does not fall into the nucleus. It gives off radiation of specific wavelengths. Bohr contended that an electron had a minimum orbit or ground state and an excited state where the electron moved by the absorption of an amount of energy. He bought Planck's constant into an equation and made the electron capable of taking on only certain orbits. Thus, the analysis of the atom quantitized.[134]

Arnold Sommerfeld (1868-1951) refined the quantum theory of the atom by suggesting that electron orbits could be elliptical. He developed the orbital quantum, the magnetic quantum number and the spin number to further expand his model. In 1925 Wolfgang Pauli (1900-1958) argued that two electrons of opposite spin would exclude other electrons. Electrons in each shell separated into sub-shells. After scientists inferred these sub-shells, the periodic table reflected the electron arrangements.[135]

Logic, Dialectic and Physics

## Energy Levels of Electrons

During the 1920's scientists began thinking not just of orbits but levels of energy. During 1925 Werner Heisenberg developed a method where he expressed energy levels as matrices. This approach helped define spectral lines produced by atomic data. In solids the idea of an energy band emerged instead of just an energy line. Even if electrons filled a bond, electrical conduction was still possible because energy absorption might move electrons to a bond of higher energy. Some substances called semiconductors easily allow this. Solid-state devices are electronic instruments that use semi-conductors. Also, a change of levels can be done by photons of a defined size creating atomic clocks and lasers. In 1953 Charles Townes subjected ammonia molecules to impact by photons of specific energy content. Thus, a few photons went in and many photons left. He amplified radiation. The word is, maser, or "microwave amplification by stimulated emission of radiation."[136]

Townes suggested the possibility of creating a light producing maser or laser. In 1960 Theodore Maiman created a laser by stimulating aluminum oxide with some chromium oxide by exposing it to light. The light created by a laser consisted of wave packets of the same frequency. These wave packets produce coherent light because they stick together.[137]

## Of Matter-Waves

Is the wave-particle duality related only to electromagnetic radiation? Louis de Broglie wrote an equation that defined wave forms on particles of matter or matter's waves. They are not electromagnetic by nature. Do electrons create matter-waves? American scientists, Clinton Davisson and Lester Germer with the Englishman, George Thomson proved this in 1927. Electrons were a smaller wavelength than visible light. Ernst Ruske used these electrons to build an electron microscope.[138]

Erwin Schrodinger applied matter-waves to atomic theory. He argued that an electron is a wave circling the nucleus. The electron is a standing wave. Different energy levels represent different waves. He developed a wave equation to analyze atomic behavior. Schrodinger invented the field of wave mechanics or quantum mechanics.[139]

In 1944 John von Neumann equated quantum mechanics and matrix mechanics. However, in 1964 Paul Dirac argued that they are not equivalent mathematically. Matrix mechanics is more accurate. Quantum mechanics can partially explain how molecules bond. Linus Pauling showed how two electrons can form a more stable relationship than they could alone. Thus, the shared-electron pool of the so-called Lewis-Langmuir model of an atom became two resonant wave forms. Resonance offers a better explanation. A wave form is harder to locate at a particular point in space. Heisenberg argued that it is not possible to locate accurately the position and momentum of a particle at once. This position is the Heisenberg uncertainty principle. Is God playing dice? Einstein questioned this.[140] Hegel's dialectics can be applied to physics models to create approximate models built from equations to test experimentally.

## Radiation

Madam Curie in 1898 labelled the penetrating and ionizing radiation coming from uranium compounds "radioactivity." She showed that various uranium compounds were radioactive. Ernest Rutherford called the three parts of radiation, Alpha rays, Beta rays and gamma rays. Antoine Becquerel found in 1899 that Beta rays coming from a radioactive substance were electrons. Gamma rays had shorter wavelengths than x-rays. In 1909 Rutherford found that the alpha particle was equal to the nucleus of the helium atom. Scientists created two devices, the Geiger counter and the Wilson cloud chamber, to study subatomic particles.[141]

In 1932 James Chadwick discovered an electrically neutral particle, the neutron. After this, Heisenberg proposed a proton-neutron model of the atomic nucleus. The neutron turned out to be an unstable particle. Scientists thought that it would breakdown into a proton and an electron. A neutron after it emitted a beta particle from the nucleus of the atom it became a proton. Thus, the atom changes. The radioactivity of uranium was a product of the change of uranium atom into some other type of atom. There was a radioactive group of elements or "daughter elements." Madame and Pierre Curie described the new elements of <u>polonium</u> and <u>radium</u>. Following this, scientists discovered some new elements.[142] Can an element have an inverse? In theory this is true.

## Isotopes

W. Prout, early in the nineteenth century, studied the atomic weights of different elements and expressed these weights in the form of the atomic weight of hydrogen. He found that integers like 4 hydrogens for helium or 12 hydrogens for carbon represented them. Later, J.J. Thomson found atoms of different mass for a different given element. He called different atomic weights for a given element, "isotopes."[143]

## Logic, Dialectic and Physics

Mentioned earlier, Ernest Rutherford did not accept Thomson's watermelon concept of the atom. He sent beams of alpha particles into thin foils of various metals. The scattering of some alpha particles hitting the metal foils was large. Rutherford thought that the positive charge and mass concentrated themselves at a point in the center of the atom. He called the heavily charged core, the atomic nucleus.[144]

Max Planck, at a meeting of the German Physical Society, proposed that light and electromagnetic radiation consisted of individual energy packets called light quanta. The amount of energy in each packet depended on its vibration frequency and is directly proportional to h, a universal constant.[145]

What is the Compton effect? It is the collisions between light quanta and electrons.[146] Niels Bohr's model of the atom had "concentric circular quantum orbits."[147] Einstein did not accept Heisenberg's uncertainty principle.[148] Gamow suggests that Dirac's positrons "can be considered...as the holes in an absolutely empty space."[149]

### A Dialectic:  Matter and Antimatter

Negative protons and positrons create antimatter. Emilio Segre was one of the first to see negative protons coming out of targets hit by 6.2 bev "atomic projectiles."[150] They created an antithesis. Also, there are quantum levels of the energy of gas particles. Thus, the quantum theory of motion impacts on the kinetic theory of heat.[151]

### Discovering The Atom

Henre Becquerel discovered radioactivity on February 27 in 1896 in Paris after a uranyle crystal darkened a photographic plate in a drawer.[152] John O'Dougherty thinks there are many accidental discoveries in science.[153] Marie Curie discovered polonium and radium after studying their radioactivity. These discoveries led to the discovery of other radioactive materials.[154] Mentioned earlier, during 1899 Ernest Rutherford found Alpha rays, Beta rays and Gamma rays. He helped discover radioactive decay with the discovery of the half life of radioactive materials.[155] Alpha particle emission represent nuclear particle decay. Beta rays are "nothing but electric adjustment of the atomic nucleus resulting from the emission of one or more alpha particles."[156]

### More on Isotopes

Logic, Dialectic and Physics

When an atom releases an alpha particle from an atom, the atom's atomic number "decreases by two and its atomic weight decreases by four." When an atom releases a beta particle, its atomic number increases by one and its atomic weight is unchanged. The release of a gamma ray leaves an atom's atomic weight unchanged. Thus, a uranium atom breakdown produces a thorium atom However, it different from a thorium atom. It differs in atomic weight but has the same atomic number. It is an isotope, a Greek expression for "same place" because it has the same position in the periodic table.[157]

Thorium isotopes have similar chemical properties. Nuclear isomers are atoms identical in atomic weight and atomic number but have different energy levels. The number of isotopes created by the radioactive breakdown of uranium and thorium form a uranium series. Isotopes have different half-lives. For example, thorium-232 has fourteen billion years half-life while thorium-231 has a half-life of about one day. It differs by a single neutron in its nucleus.[158]

In 1912, J.J. Thompson and in 1919 Frances Aston developed devices for checking rays. These rays made of ions of a given mass focused at a point or a succession of points, "(a mass spectrum)J." From the location of the points one could discern the mass of the particular isotopes and from the darkness of points on the photograph their frequency. They named the instrument the mass spectrograph. Using this device, a table of isotopes and their weights and abundance could be drawn up.[159]

### Of Nuclear Chemistry

The term "atomic weight" refers to the weighted average of isotope masses found in their distribution within an element. <u>Mass number</u> is the relative mass of a particular isotope. Isotopes partake in radioactive dating.[160]

Can the particles in a nucleus change position? In 1919 Rutherford found a "proton had been knocked out the nitrogen nucleus by the alpha particle." This was the first nuclear reaction created by a man. An electron accelerated by a field of electric potential of one volt develops an energy of one <u>electron-volt</u>. Electron-volts can define mass. The mass of an electron equals 510,000 electron-volts. Also, electron-volts define wavelengths of electromagnetic radiation. To create subatomic particles with high energies, scientists built a particle accelerator. John Cockercroft and Ernest Walton built the first successful accelerator that moved protons in 1929. This device used condensers to increase potentials to high levels. During 1931 they accelerated protons to disrupt a nucleus of lithium.[161]

In 1931 three accelerators appeared. First, was the Robert Van de Graaf an electrostatic generator that produced a potential difference by moving a belt in half a dumbbell. Second, scientists created a linear accelerator that gave several potential kicks to the particles in a line tube. It grew to be too long. Third, Ernest Lawrence (1901-1958) built a curved "cyclotron." However, increases in mass built in limits on the size of energies placed on a proton. In 1945, Edwin McMillan and Vladimir Veksler showed how an alternation of potential decreases to keep in sync with more massive particle. To further deal with this problem, Donald Kerst built a betatron that increased the magnetic field as electrons moved faster in a circle.[162]

### The Creation of Artificial Radioactivity

Frederic and Irene Joliot-Curie were the son-in-law and daughter of Marie and Pierre Curie. They continued Rutherford's alpha particle bombardment of different nuclei. Frederic and Irene Joliot-Curie created in the lab phosphorus-30 an initial isotope that did not naturally occur in the earth. It was the first creation of artificial radioactivity. Over the next twenty years scientists invented around one thousand radioisotopes . Gradually physicists invented chemicals containing these isotopes, for example, stable hydrogen-2. In 1935 a biochemist, Rudolf Schoenheimer, used isotopic tracers to study tissue reactions.[163]

After scientists discovered the neutron, they used it as a bombarding particle to create nuclear reactions. The neutron lacked a charge. Thus, it was difficult to accelerate. In 1935 John R. Oppenheimer suggested the use of deuteron which contains a proton and a neutron in a loose combination. Enrico Fermi found that neutrons created nuclear reactions more effectively after passing through paraffin or water. The size of the target hit by a single cross section is the "nuclear cross section." After methods arose that created isotopes nuclear reactions, scientists developed elements not appearing in nature, such as francium and einsteinium.[164]

### Results Regarding Nuclear Structure

Hydrogen-1 is an atom containing a single proton. However, if any nucleus contains more than one proton, it also contains a neutron. Many nuclei of high mass numbers contain more neutrons than they contain protons. The neutrons help stabilize the nucleus. When a uranium nucleus absorbs a neutron under bombardment it might break in half in a process called nuclear fission. This process led to the creation of nuclear reactors. Lise Meitner, Paul Weger and Edward Teller worked to control a chain reaction that uranium fission could produce. They worked to purify uranium and separate its isotopes.[165]

At the end of 1942 Enrico Fermi tried to create a self-sustaining nuclear reaction at the University of Chicago. The first nuclear reactor weighed 1400 tons. It contained 52 tons of uranium. On December 2, 1942, at 3:45 p.m. Fermi partially pulled out cadmium rods and in so doing produced a self-sustaining reaction. The cadmium rods absorbed neutrons. By pulling these rods out more neutrons were available to cause uranium atoms to fission. The creation of the atomic bomb, nuclear submarines, nuclear ships, and breeder reactors soon followed.[166]

Scientists developed during 1945 three fission bombs. The first they exploded at Alamogordo, New Mexico on July 16, 1945, at 5:20 A.M. On August 6, 1945, the second atomic bomb incinerated Hiroshima, and on August 8 the third atomic bomb wiped out Nagasaki.[167]

What is a breeder reaction? For example, when neutrons emerge from a U-235 core bombarded a shell of uranium covering the core, amounts of uranium-238 become plutonium-239. As a result, the amount fissionable material created in the shell may be greater than the amount consumed in the core. In nuclear fission simple atoms are built into more complex ones. At very high temperatures, atomic nuclei are forced together and thermo nuclear reactions arise. Today fusion bombs with over 50 million tons of TNT have been created. Fusion reactions can power the earth. Moreover, the field of plasma physics deals with atom fragments called "plasma."[168]

### Of Anti-Particles

Robert Millikan discovered in 1925 a new type of radiation called "cosmic rays" coming from the cosmos. During the 1930's A.H. Compton proved that cosmic rays contained particles and were not electromagnetic. Thomas Johnson found that the particles had positive charges.[169]

In the 1940's high-altitude balloons and rockets studied <u>primary radiation</u> or the cosmic particles that had not collided with nuclei. Later, Paul Dirac predicted the existence of the positron. In 1932 Carl Anderson working with a cloud chamber discovered a positively charged electron or positron. Today scientists classify the positron with a class called anti-particles. A positron and electron together cancel each others charge or mass in matter annihilation. A

Logic, Dialectic and Physics

positronium is an atom that contains an electron or a positron circling each
other around the same center of gravity.[170]

In 1956 Owen Chamberlain and Emilio Segre discovered an antiproton.
Perhaps anti-particles can produce "anti-gravity." Anti-matter contains
antiprotons, antineutrons and positrons. Pauli predicted that the neutron, the
proton, the electron and the neutrino or ("little neutral one") exist. If neutrinos
exist, then scientists can search for antineutrinos. Later Hideki Yukawa and
Carl Anderson discovered the muon or "heavy electron." It may orbit within
the nucleus.[171]

During 1947 Cecil Powell discovered the pi-meson or pion with a mass
equal to that of 237 electrons almost the mass of Yukawa's so-called
"exchange particle of the nuclear force." Today scientists have found some
particles, for example, the K-mesons which equal about 966.5 electrons in
mass and lie between protons and pions. Tsung-Dao Lee (1926) and Chen
Ning Yang (1922-) found that the law of conservation of parity need not
always be true for weak interactions. It held for strong interactions.[172]

Is relativity an absolute? Is the speed of light an absolute? The answer
is that scientists qualified Einstein's theories. Scientists built approximate
models to Einstein's theories.

In sum, there are physics and chemistry problems that elements of logic
like the antithesis or the inverse can explain. For example, duality is a major
theme in physics. It occurs in the atom, in magnets, in gravity, in theories
about light, and in a potential unified field theory synthesis. Thus, logic
extends to matter.

Logic, Dialectic and Physics

## Physics, Dialectics and Oriental Culture

Fritjof Capra contends that two major themes of Eastern thought are the unity of all phenomenon and the dynamism of the universe. Eastern mystics perceive the world as interacting and unified, with a person as an actual part of this organic system. These mystics often see physics as unimaginative and the originator of many of the evils of technology.[1] Capra ignores the East's orientation to reincarnation and atheism.

Dialectics are key to physics and to oriental mysticism and reincarnation. The "birth, death, and rebirth" of the universe are examples of dialectics and are the basic laws of nature.[2] Fritjof Capra will die and be reborn again in the next world. "The earth is a special place" in the universe thinks James O'Dougherty.[3] "Language, is the damnation of the orient" thinks John O'Dougherty.[4] Maybe this is true for their mathematics. Margaret O'Dougherty points out that the French language labels all the nouns with masculine and feminine referents. This is also true of some Egyptian language referents.[5] In a sense then these languages are more dialectical than the English language. Oriental science is a thought disorder thinks the writer.

Capra suggests that the mind has two types of knowledge: the rational and the intuitive. Eastern mystics favor intuitive knowledge. Abstraction is an integral part of rational knowledge. Science is an extension of rational knowledge. However, Werner Heisenberg contends that every word has only a small range of applicability. Eastern mystics focus on that part of the experience of reality that transcends thinking and the senses. Thus absolute knowledge is an intellectual experience of the universe. In contrast, physics bases its theories on experiments and mathematics. However, many mathematicians think that mathematics is a part of nature. Pythagoras said, "All things are numbers." Western religious ideas often combine mathematics and theology.[6]

In the East mathematics is like a "conceptual map" and not a feature of reality. The mystics parallel scientific experiments and various mystical experiences. Some Eastern schools like the Hindu Vedanta value the intellect. In contrast, Taoists mistrust reason. Mystical teachers often emphasize the theme that experience is repeatable like an experiment.[7]

What is the main element in Eastern meditation? It is the experience of oneness with nature. Karate is a form of meditation in the East. The field of physics, in contrast, uses models. They are appropriate for only a delimited range of phenomena. Eastern mystics often argue that myth is the way to

55

bring a person to absolute truth. Eastern poets are aware of the limitations of language.[8]

## Language and Beyond

Language is inadequate to define atomic and subatomic reality. In the East it is thought reality transcends everyday language. For example, the <u>Tao Te Ching</u> is paradoxical by nature. Today quantum theory and relativity have led us to a holistic and "organic" orientation towards nature. Newton's physics stated the ideas of an absolute space and time, a causal depiction of phenomena, and an objective approach to nature. Einstein developed an intellectual framework which he applied to mechanics: "the special theory of relativity." He argued that space has more than three-dimensions. It relates to time to build a four-dimensional continuum.[9]

During 1905 Einstein published his general theory of relativity in which he applied the special theory to gravity. Einstein argued that three-dimensional space curves and that this curvature is the result of the gravitational fields of large bodies. Time affects matter. Space-time relates to matter's distribution in the universe and "empty-space" loses much of its meaning.[10]

Max Planck discovered that heat radiation emits in "energy packets" or "quanta." Einstein suggested that electromagnetic radiation and light appear not only as waves but as quanta. Particles have wave properties. These waves are not like three-dimensional waves; instead, they are "probability waves." Matter is not isolated but contains a web of relations. The Cartesian wall between the I and the universe is not useful in dealing with atomic matter. Today in physics we cannot talk about nature without including men and women in the equation. Also, electrons form "standing wave" patterns like music waves. "Quantum numbers" which indicate their states define their orbits. Besides electrons scientists have found over two hundred particles.[11]

Paul Dirac wrote a relativistic equation explaining electrons. He contended there is a symmetry between matter and antimatter. Dirac predicted the existence of the positron which has the same mass as an electron but an opposite charge. For every particle an antiparticle exists with an equal mass but opposite charge. There is symmetry. Material particles can be created from "pure energy" if enough energy is used. Particles can be created and annihilated. To divide subatomic particles we collide them. Matter divides over and over. However, we do not obtain smaller pieces because we create particles out of the energy used in the process. Subatomic particles are thus both destructible and indestructible at once. The idea of an "elementary particle" loses its meaning. Particles dynamically interact with their environment.[12]

In relativity, scientists depict the forces between particles as the exchange of different particles. Force and matter link. Moreover, forces of matter relate to other constituents of matter. Thus, the forces between constituent particles and particles composing the binding forces blur. Space, time, and cause and effect have lost much of their meaning. The universe is a dynamic whole.[13] An approach to relativity is that life is relative.

### Of Hinduism

Hinduism is not a well defined philosophy or religion. Instead, it is a complex socioreligious organism that contains several sects, cults, gods and goddesses. Its focus is India. The spiritual fount of Hinduism is the Vedas. There are four Vedas and each contains sacred hymns and prayers. Other parts reveal sacrificial rituals and hymns. The Upanishads Veda develops the philosophical context of the early Vedas. The Indian people, however, relate to Hinduism through epics and poems like the Bhagavad Gita.[14]

The basic idea behind Hinduism is that true reality or Brahman exists in the variety of things around us. This Brahman is the 'soul' or essence of things. It is incomprehensible and has no beginning and is unthinkable. In Hindu thought God takes on the world and the world takes on God. The creation play of God is lila. Maya is the term that refers to the psychological state of a person who has magic. Maya changes a lot and in nature "all forms are relative."[15]

What is karma? It is the life force. To free oneself from the magic of maya is to free oneself of the bonds of karma or to see that all sensory phenomena is a part of the same reality. To experience personally this everything including oneself is Brahman. It is moksha and it is the essence of Hinduism. Brahman is beyond images and concepts.[16]

The Vedanta school which is a product of the Upanishads see the Brahman as a metaphysical idea free from myth. In contrast, to many schools of Western philosophy these mystics meditate. Yoga is the joining of the personal soul to Brahman.[17]

Most Hindus worship a god or goddess like Shiva, Vishnu, or the Divine Mother. Shiva is the deity of destruction and creation. Through dance, Shiva maintains the rhythms of the universe. Vishnu is the preserver of the world.

Logic, Dialectic and Physics

The Divine Mother, Shakti, contains the universe's feminine energy. Sometimes Shakti is Shiva's wife. Hindus do not suppress sexual love. It is a part of the spirit. All of these divinities have the same substance and all reflect Brahman.[18]

## Of Buddhism

Buddhism, in contrast to Hinduism, traces its roots to Siddhartha Gautama or Buddha. He lived in India during the sixth century B.C. Buddhism is largely a psychological philosophy. The key to psychology is dialectics. There is a right and left brain. Buddha concerned himself not with the Divine but the human. In a way he was a psychotherapist. Buddha gave the Indian ideas of maya, karma, and nirvana a psychological emphasis.[19]

The Hinayana and the Mahayana schools of thought arose after Buddha died. The Hinayana is the orthodox school. In contrast, the Mahayana is more flexible. Both of these philosophies spread across Asia. Mahayana Buddhism shuns speculative thought. It looks towards mystical experience or 'awakening.' The apex of experience according to this philosophy is the acintya or unthinkable. At this stage reality does not differentiate. Buddha, himself, became the 'awakened.'[20]

Buddha preached four main truths. The first truth of man is suffering. Life is transitory. There is no self. The second truth is the cause of suffering, namely clinging and grasping. The wrong point of view grasped by man is ignorance or dividing the world into categories. Man makes the mistake of clinging to categories or groups. Clinging to objects is samsara or the cycle of birth and death. Karma the chain of cause and effect fuels it.[21]

The third truth is to free oneself from karma and to reach liberation or nirvana where the oneness of life become a persistent sensation. The fourth truth is the eightpart path of development which brings one to Buddhahood. The first two areas of this path deal with right knowing and seeing. The next four areas deal with right action. They emphasize the means between the extremes in life. The last two areas deal with meditation and mystical experiences. Buddha did not play up authority. Each person he thought must find his or her own path.[22]

Nagarjuna, a Mahayana thinker, argued that reality is not just concepts. He thought essence is emptiness. However, Buddhism also emphasizes "faith, love and compassion." It stresses two other elements, intuitive intelligence and love. One finds the culmination of Buddhist thought in the Avatamsaka which sets forth the themes of unity and the interrelation of all objects and all life. This theme parallels some theories in modern physics.[23] **Not all theories relate. There are many paradoxes in language and science. Hegel's philosophy is paradoxical.**

## Chinese Philosophy

The Ancient Chinese philosophy had two sides. One side emphasized social consciousness and the other side emphasized the unity of the sage with the universe. The sage concerns himself with both the spirit and the world. He finds unity by bringing together intuitive wisdom and social action.[24]

In the sixth century B.C. two different schools arose, Confucianism and Taoism. Confucianism stressed education, social organization, common sense and practical knowledge. It built an ethical basis for the Chinese family. In contrast, Taoism dealt with the study of nature and its Way. Human happiness occurs when one follows nature.[25] Confucius based his Six Classics on poetry, music, and the "holy sages" in Chinese history. He was not the editor of these books but his disciples passed on his wisdom.[26]

Lao Tzu started Taoism and wrote a book called the Tao Te Ching. The work contains many paradoxes and much poetic language. A second Taoist book is the Chuang Tzu. The Chinese thought, like many Indian thinkers, there is a unifying reality which underlies all the events of life. The highest reality is the Tao. Reality is undefinable. The world is constantly changing. The principle theme of the Tao is the cyclic nature of change. Not all numbers are cyclic. Many chemical reactions are irreversible. The mean between the extremes or the golden mean is a major theme of Taoism and Confucianism.[27] There are extremes in nature, for example, absolute zero. Reincarnation and atheism are extreme views of life.

In Taoism the yin and the yang are the poles that set limits on change. Yin is the female power linked to earth and Yang is the male power linked to heaven. Rest is yin and movement is yang. Yin is the feminine intuitive mind and yang is the male rational intellect. Life is the harmony of yin and yang. The idea of the neuter is an important part of science and life in the English language.

The Confucian Classic, I Ching or Book of Changes, contains some hexagrams either broken yin or unbroken yang and several archetypes representing the Way. The major theme of the I Ching is dynamism or change. The universe constantly transforms.[28]

Logic, Dialectic and Physics

## Of Taoism

Taoism deals with intuitive wisdom as opposed to rational knowledge. Taoists think there are limits on reason. Hegel sees, in contrast, the strength of reason or pan-reason. Taoists see change as an essential part of nature. Polar opposites like the yin and yang are relate to each other. In human conduct they think that if you want to reach a goal you should begin with its opposite or its antithesis. Good and evil interrelate. Moral standards are largely relative. Heraclitus, the ancient Greek thinker, also thought there was continuous change in nature and that opposites unite. Taoists think change is not just the result of some force but and innate tendency.[29] In contrast, Christians think that love is an absolute. James Joyce thought "there will always be love."

## The Philosophy of Zen

The Japanese took on Zen philosophy around 1200 A.D. It is Buddhistic and its goal is enlightenment called satori in Zen. Satori transcends thought categories. The ultimate truth is beyond language. Zen teachers point to the truth with riddles and paradoxes. Satori is the immediate experience of the essence of all objects especially the experiences of everyday life. Zen followers try to live life spontaneously. The two schools of Zen, the Sato and the Rinzai schools, give importance to sitting meditation which is done in Zen monasteries daily. Zen followers maintain that enlightenment occurs in everyday life, for example, in painting, judo and the tea ceremony. All of these activities lead a person towards enlightenment. Zen mastery transcends art and becomes "artless" developing out of the unconscious.[30]

## Parallels between the Tao and Science

The basic theme of the East is the unity and interrelation of all realities or oneness. Hegel, also, emphasizes monism. Capra feels that this is an important theme of modern physics. In quantum theory, systems are only probable. Electrons have patterns of interactions in regions. The idea of a seen object has flaws. Quantum theory highlights the interconnectedness of all realities. The universe is a web. Scientists study in atoms "objects" which one discovers largely in measurement and analysis. Science is an interplay between nature and humanity.[31] There are research parameters in any scientific experiment. Scientists make many errors. If all objects are interconnected why is the earth so "special?"

## Logic, Dialectic and Physics

The idea that good and evil, dark and light are different realities of the same phenomena is a principle of Eastern life. Balance is also fundamental to the unity of opposites. Polar opposites have dynamic unity. Today in physics subatomic particles are both continuous and discontinuous. Heisenberg's uncertainty principle highlights the uncertainties between the position and the momentum of particles. A particle's "wave packet" is a vibrating pattern in time.[32]

## Space and Time

Einstein's central thesis is that geometry is not inherent in the universe. It is a construct of the mind. Capra thinks that Eastern philosophy has always thought that space and time are constructs. Einstein contended that spatial and temporal definitions are relative. Space and time interpenetrate each other. Einstein theorized that the gravitational fields of large bodies curve space. This distortion also affects time. In physics particle interactions form a "four-dimensional snapshot" dealing with all time and all space. For each process there is an equivalent process with time reversed and "particles replaced by antiparticles." On a sphere there can exist for a triangle three right angles. Black holes exist where a star collapses due to gravitational attraction of particles and creates a space around a star which curves so that light does not come to us.[33] God has a subtle mind. Capra ignores reincarnation and atheism as keys to the oriental mind. These are anti-Christian dogmas.

Eastern mystics try to transcend time: "In the absolute there is neither time, space, nor causation." In the world "time, space, and causation are like glass through which the absolute is seen." Maybe as Julie Bates thinks "darkness can come out of light."[34]

## The Dynamism of the Universe

The dynamic nature of the universe is central to Eastern mystic thought. The Hindus call the reality of the universe, Brahman, the Buddhists "the Body of Being" or Suchness, and the Taoists, the way or Tao. The word Brahman shows maturation. The Vedic word karma indicates "action." Buddhism emphasizes the impermanent. In the Tao reality is constantly changing. Present day physics depicts matter as vibrating and rhythmic.[35]

In space stars explode, spin or contract. The universe is expanding. A "big bang" created the universe millions of years ago. Some scientists think that the universe will contract and expand again. Birth, death, and reinvention are the keys to the universe. We can reinvent a dead star. The Hindus, depict a universe that expands and contracts. Einstein's equivalence of mass and energy predicts that the collision of particles can transform the masses and kinetic energies of newly fashioned particles. The total energy remains the same. Quantum theory shows that particles are patterns of probabilities interconnected in an inseparable web. Matter and activity are different manifestations of space-time reality. Eastern mystics contend that the elements are constantly undergoing change.[36]

### The Void and Form

After the emergence of field theory, physicists tried to incorporate various fields in physics into a new synthesis or a unified field. Eastern mystics have emphasized the unity of all phenomena. They think that reality is formless or a void. It is this void which gives birth to the world's forms. In quantum physics the field is a continuum which is everywhere in a space but has a discontinuous structure. There is a dynamic unity of opposites. In Eastern mysticism there is a unity between the Void and its forms.[37]The Catholic Church teaches that "in the beginning was the word."

In electromagnetics a field can express itself as waves or photons or as a field between charged particles. For example, a force can express itself as an exchange of photons between particles. An electric repulsion can mediate photon exchanges. It is an intersection rather than a force. A neutron can emit and reabsorb a pion. Also, nucleons can emit and reabsorb particles often. In nuclear interaction particles and forces relate to their "virtual clouds," for example, "virtual mesons." Thus, particles have dynamic patterns. There is not the ancient distinction between particles and empty space.[38]

### Of Mesons and Hyperons

Hidekei Yukawa during 1932 suggested that nuclear cohesive forces were due to a new particle, the meson, exchanged between protons and neutron like dogs exchange a bone (this example is Gamow's). Later, scientists found a heavier particle or muon with a mass of 273 compared to the mesons mass of 206.[39]

### Parity or Mirror-Symmetry

Chen Ning Yang and Tsung Dao Lee, during 1956, found that in a beta-decaying radioactive substance all the electrons emitted in the same direction. Equal numbers of them did not fly "towards the north and south poles of the electro-magnet." Thus there was not mirror-symmetry in subatomic physics. In regular physics this symmetry occurs.[40]

## A Dance in the Universe

Particle interactions build stable structures which are not static but oscillate in dance like rhythms. These patterns form groups. There is a lot of order. All atoms contain three large particles: "the proton, the neutron, and the electron." Another particle, the photon, has no mass and stands for a unit of electromagnetic radiation. The neutron can decay spontaneously. It becomes a proton, an electron, and a neutrino. Neutrons lack mass and electric charge. Each particle has an antiparticle with the same mass but reversed charge. For example, in beta decay an antineutrino emerges.[41]

Today there are many known subatomic particles. Most have short life spans. Death is a small particle reality. Scientists create and destroy each of these particles in collision processes. Each particle can be exchanged and in so doing give to the particle's interaction. Capra contends that these interactions fall into four groups with different strengths:

(1) strong interactions
(2) electromagnetic interactions
(3) weak interactions
(4) and gravitational interactions.

Neutrons and protons hold together by nuclear forces of about ten million units. Electromagnetic interactions occur between charged particles. Gravitation interaction occurs between all particles but is very weak.[42]

Of the subatomic particles only some do not take part in strong reactions: the photon and the "leptons." "Hadrons" participate in strong interactions and divide into "mesons" and "baryons." The baryons have antiparticles. The meson is its own antiparticle.[43] A boson is a particle that does not spin.[44]

The leptons take part in weak interactions. These interactions occur in special particle collisions and particle decay like beta decay. Hadron interactions occur with the exchange of different hadrons. The exchange of these particles creates strong reactions. Their range is only a few particles in size and they never develop a macroscopic force. In contrast, electromagnetic interactions exchange massless photons and have a long range. Scientists think gravitational interactions  involve a massless particle, the "graviton." Gravitons have not been seen yet.[45]

Weak interactions have short duration and scientists think they involve the interplay of a heavy particle, the "W-meson."  Scientists think its role is similar to the photon's role in electromagnetic interactions.  However, it has a larger mass.  Particles destroy and recreate in the collisions and the interactions of high-energy physics.    Thus, "birth, death, and rebirth" are the keys to matter.  In the center of stars subatomic particles appear frequently.  There, high energy collisions occur often.  These collision processes often create strong electromagnetic radiation like radio waves, light waves, or x-rays. They fill the galaxies with photons, that is, electromagnetic radiation.  "Cosmic radiation" contains photons and other massive particles whose source is unknown.  These rays create a "cosmic dance" in the atmosphere.[46]

Particle exchanges are complex.  For example, a proton can emit and reabsorb "virtual" pions constantly.  A negative pion can produce a neutron and an antiproton which annihilate each other to rebuild the original pion.  In Hindu culture there is a dancing god, Shiva, whose dance symbolizes creation and destruction cycles and life's daily rhythms.  She reveals to us that the forms in the world are always changing.[47]   Kepler thought that the sphere of the universe had music.

## Of Quarks

All particles of a given type are the same.  For example, protons have a given mass and electrical charge.  Particles also have similar patterns.  Many spin and their spins have definite values.  Hadrons, for example, have well defined "resonances".  They can reabsorb energy to develop a variety of "excited" patterns.  Subatomic particles form processes and interactions.[48]

The idea of symmetry, for example, mirror reflections and the yin and the yang of weak interactions help classify particles.  Interactions of particles are symmetrical in space, time, rotation, and electrical charge.  Scientists use quantum numbers to identify and situate particles like baryons in symmetrical families of patterns.  They think hadrons contain of smaller entities called "quarks." They have special symmetries somewhat like elements of the Zen Koan.[49]

## Designs of Change

## Logic, Dialectic and Physics

In 1943 Heisenberg conceptualized the "S-matrix" to explain strong interactions like hadron reactions. Hadron reactions depict an array called a matrix. The S stands for "scattering." This matrix is different from Feynman's field theory diagrams. Also, these matrices are not space-time diagrams. Instead, they are symbolic presentations of particle interactions. Reactions define velocities of particles coming in and going out. Thus, a S-diagram contains less information than a Feynman diagram. To deal with the Heisenberg uncertainty principle the S-matrix specifies particle momentum which remains imprecise about the region of the interactions. The S-matrix does not look mainly at particles but reactions of particles. This matrix combines quantum mechanics from objects to events with relativity's conception of particles in space-time.[50]

Hadron reactions stand for an energy flow which occurs in "channels formalized by quantum numbers." The chance of two hadrons interacting in a collision is a function of the energy involved. These hadron states are "resonances." The S-matrix theory is the first movement towards a dynamic explanation of the patterns that particles form. "Channels" are different aspects of reactions and the crossing of channels are the fine points of the S-matrix approach.[51]

Three general principles are the key to this approach. First, relativity states that reaction probabilities are independent of space and time in an experiment, independent of the experiments orientation in space, and independent of the observers motion. Second, quantum theory predicts that the probable outcome of the interaction of particles will be equal to one. That is, they will either interact or not. Third, is the principle of causality. Energy transfer occurs where a particle created in one reaction happens only if the "latter reaction occurs after the former." When new particles arise the S-matrix makes a singularity. The matrix does not determine the singularity.[52]

In Chinese thought, hexagrams depict human situations like "the Receptive." Thus, the Chinese have an idea of change like the S-matrix theory. In S-matrix theory processes like particle reactions form the world of hadrons. In the I Ching these processes depict "the changes" or inner tendencies. The reaction probabilities in S-matrix mathematics also reflect tendencies within.[53]

### Of Interpenetration

The Eastern Mystics agree with the idea of the so-called "bootstrap philosophy" that the universe is a connected whole where no point is more basic than any other. The properties of one determine many of the rest. Each part "contains" all the others. The universe is indivisible. The Chinese do not have a phrase that relates to the Western concept of a "law of nature." Buddhists try to get at 'absolute' knowledge that involves the sum of all life.[54] Life is not just matter it is the spirit. Hegel thinks that spirit interpenetrates the universe.

The bootstrap idea stands for a view of the universe that grew up in quantum theory with a realization of a universal interrelationship, a dynamism found in relativity and S-matrix probabilities. The bootstrap theory denies the basic consistency of matter, the fundamental laws of equations, and the laws of nature--the foundation of Judaeo-Christian traditions.[55] However, Terence O'Dougherty, David Noble think that death is a law of nature.[56] Does it deny death? David Noble thinks that the basic reality of the universe is "birth, death, and rebirth."[57] This is the Western key to oriental thought.

In conclusion, there is enough mass in the universe for it to collapse, contract, and be reborn. David Noble thinks that "the earth will die, the universe will die. We all will die."[58] This is a fact of life. God is interpenetration. Physics is reinvented through logic. James O'Dougherty and David Noble think the earth is a special place. It is interpenetrated with the Spirit.[59]

Logic, Dialectic and Physics

The Spirit of a Physics Department

Part II The History of University of Minnesota's Physics
Department
Thesis: Intellectual History, Philosophy and Physics are Equivalent Fields

## A Conceptual Framework from Robert Solomon

Hegel wrote his Phenomenology as a preface to a greater philosophical
work which starts with The Science of Logic and continues with his
Encyclopedia of the Philosophical Sciences. He called it a "bacchanalian revel."
Solomon argues that Hegel was a humanist. He views God as the human spirit
or Geist. Solomon continues and says that Hegel was an "anti-Epistemologist."
Hegel is a "phenomenologist" which meant to Hegel the systematic depiction
of experience. The context of consciousness makes experience "necessary."
Ideas deal with time and are not just products of time. Reason is "the search
for unity." Hegel is a philosopher of change and of the absolute or of a unified
world view. Dialectics depict changes.[1]

## Hegel and Politics

The French Revolution occurred while Hegel was growing up. Hegel's
focus in the Phenomenology was to show the chances of unity in the many
cultures and a world-view of humankind. Hegel emphasizes unity. Ironically,
his Enlightenment thinking depicts humanity: morality, politics, reason and
religion. This tension splits the world of phenomenology. Hegel did not have
a single political outlook argues Solomon. He concerns himself with freedom.
Politically, Hegel's idea of freedom is both "freedom from" and "freedom to."
Freedom is the self-realization of identity.[2]

## Dialectics

Solomon argues that in the Phenomenology the term "dialectic" is more
like a metaphor for "growth" or "metamorphosis." The universe is like a "living
process." The history of physics then progresses as our consciousness
develops. Plato thought that dialectics are a means to find truth where Kant
thought that dialectical truth proves that truth is not beyond "phenomena."
Hegel's dialectic combines these approaches. Like Kant, he thinks that reason
creates contradictions. Like Plato, he feels that they are a clue to truth. As
we see inconsistencies our consciousness and comprehension develops. The
end of this process is "absolute knowing" or the creation of a single conception
that makes sense of all immediately. Solomon argues there is an existential

side to Hegel. We create our meaning in life and we are responsible for our world.[3]

## Time and Spirit

Hegel thought that philosophy expressed culture and time. Solomon suggests that Hegel interests himself in the perfection of humanity or spirit. Hegel was a product of the French Revolution and Napoleon argues Solomon. Hegel used the phrase "the universe actualized through the particular" to reflect the creation of Germany's self-identity. Culture, for Hegel, is identity. Individualism is nothing. Athenian culture influenced him. The German nationalists excited Hegel.[4]

Solomon suggests that the key metaphor to Hegel's phenomenology is the German word Bildung meaning semi-biological development and **"growth."** The goal of Bildung is rational self-realization. Hegel favored such metaphors as "progress, redemption, mission, and humanity." Did Hegel wish to become a Protestant Aquinas? Perhaps he is. He blended Christianity and humanism.[5] Ideas are the bootstrap of the universe.

## Kant and Hegel

Hegel related to Kant's philosophy. Kant synthesized Leibniz and Hume or "rationalism" and "empiricism." Hegel and Kant were idealists. That is, the world is to a certain degree a product of consciousness. Christians, for example, look at the spirit within.[6]

Kant tried to reconcile scientific knowledge, practical morality, religion, Newton's physics and Christianity. His Critique of Pure Reason tried to defend "the metaphysical principles of science" and knowledge. Kant's Critique of Practical Reason defended the ideals of morality and Christianity. His Critique of Judgment tried to unite the previous Critiques and defined and defended the aesthetics of taste and the "Spirit." There is a priori knowledge. This knowledge starts with experience but does not always arise from experience. Space and time, for example, are a priori forms of thinking. "Kant synthesized British empiricism and Continental rationality," argues John Harris.[7]

Kant invented the dialectic and four antinomies:

      I.    **Thesis: The world has a beginning in space and time. Antithesis: The world does not have a beginning in space and time.**

Logic, Dialectic and Physics

   II. Thesis: Everything consists of simple elements. Antithesis: There are no simple elements.

   III. Thesis: There are causes through freedom. Antithesis: There are no causes through freedom, only natural causes.

   IV. Thesis: There is a necessary being. Antithesis: There is no necessary being.[8]

  It is possible to synthesize these dialectics. First, the world always existed in God's mind and it evolved during creation. Second, there are both simple and complex elements. Third, there are both free causes and natural causes. For example, a person has a will and an autonomic nervous system. Finally, God is a necessary being but man is not.

  Kant was a dualist. First, each person has a transcendental self which applies classes of understanding to the universe of experience. Second, each person has a rational, willing, and acting self which stands with God beside time and space.[9]

## Hegel in His Youth

  Hegel was an abstract defender of the idea of the absolute. Solomon argues that the Enlightenment influenced Hegel. For example, he accepted religious tolerance, Rousseau's emphasis on the significance of education, the perfectibility of man, and the goodness or man's nature, and the idea of alienation. Hegel was anti-Catholic and a nationalist. He entered the Tubingen Theological Seminary when he was seventeen.[10]

  Hegel accepted Rousseau's idea of inborn potential. He thought that religion should appeal not just to the intellect but also to man's happiness, to virtue, and to "the heart." Reason is the basis of subjective religions. Hegel emphasized community rather than individualism. The basis of religion is community. The "subjective" is universal. Hegel localizes objectivity. Religion is not as Kant and later Kierkegaard thought "merely personal." Hegel's early writings emphasize "growth" and inborn virtue. Philosophy should have vitality. Hegel favors German Protestantism. He does not ignore Christianity; instead, it pervades his philosophy. In his "Life of Jesus" Hegel secularizes the New Testament. He remained ambivalent to Christianity and Jesus for his whole life.[11]

**Logic, Dialectic and Physics**

Hegel's <u>Phenomenology of Spirit</u> suggests "the absolute" contains all "forms of consciousness." He attacks religion for its failures in society. An idea of "natural" law he finds in humanity.[12]

How could a "slave religion" like Christianity dominate superior pagan religions? Solomon thinks that Hegel interprets Jesus as being somewhat proto-Kantian. Hegel looks at the female side of truth and at <u>Aether</u>, the substance that binds together the universe. He thought that a Greek folk-religion might supersede Christianity. Hegel emphasizes the "spirit" meaning something akin to "inner unity."[13]

Hegel argues against empiricism and the "metaphysics of subjectivity." Infinity is "immanent." Eternity and the Eternal Idea express the world's logical changes. Thought is both subjective and objective. Natural law is an extension of reason.[14]

Hegel's, <u>Phenomenology</u>, studied the structure of the world's "phenomena." He followed up this work with a study of "practical reason" with a focus on "God, freedom and immortality. "The absolute" is ground in the concrete problems and questions of life. Hegel argues for "the eternity of truth, the unity of reason, the essence of philosophy. A process is a unity. Concepts are fluid. Different languages give different meanings to concepts. The end of philosophy and religion is, Hegel thought, truth.[15]

Hegel's idealism contends that objects are dependent on God's mind. He rejects the distinction between mind and matter, consciousness and object, and experience and reality. Hegel's "the absolute" refers to reality and God. Hegel understands the absolute by the "Concept." He contends that the world is contradictory but reason can reconcile the contradictions. The Spirit and the Absolute are identical.[16]

Hegel's one consciousness, Spirit, includes the sum of human life: Consciousness and the universe. "The <u>Phenomenology</u> is a proof and the dialectic is its method. How does Hegel view deduction? He does not start with a "first principle." This ends the essence of "deduction." Deduction is to the ends that desires are to the real. It is not the a following of the rules of logic. It is a journey and a conversation rather than a logical proof. The end of philosophy and the Spirit is total comprehension. Truth and meaning are the whole or <u>holism</u>.[17]

There are three main sections to the <u>Phenomenology</u>--(1) Consciousness, (2) Self-Consciousness and (3) Reason. Consciousness and self-consciousness

Logic, Dialectic and Physics

unite in reason. Hegel thinks that the principle of consciousness is the comprehension of the experience of the world. The Phenomology's "logic" is a progression of representations of life or approximations. It is a journey rather than a tennis match.[18]

In his Preface Hegel mixes epistemological and biological metaphors. Truth grows. The shape of truth exists in the scientific system. Science means in this case the necessary propositions as opposed to empirical results. Hegel presents the idea of the World Spirit. The Spirit is not complete. The key part of truth is the concept. We reach the Absolute by conceptualization.[19]

What is the truth? It is those concepts which we experience and form the structure of the world. The Phenomenology lacks a "method." Instead, the readers follow the transforming Concept. Consciousness transforms itself. The dialectic "is the truth." The ingredients of the dialectic are propositions, concepts and languages.[20]

Hegel thinks that concepts help us to understand experience and help us form our experience. Consciousness and objects are not completely separate. "Consciousness," "self-consciousness" and reason are different "levels" of consciousness. All three levels are experience.[21] Concepts are the key to the history of science.

Realism is the philosophy that the world is "there," a given presence. What is the disagreement on what is there? Hegel attacks "Sense-Certainty." He contends that thought and language are significant to perception. Little is given and "Sense-Certainty" cuts across many philosophies. The position of "sense-certainty" is that knowledge is "immediate" not mediated by ideas or concepts. Hegel thinks "there are no uninterpreted experiences." Nietzsche says "there are no facts, only interpretations." This point of view applies to scientific facts especially facts in physics. Hegel suggests that the key to sense certainty is the universal. He thinks that truth is "nothing in particular."[22] Hegel fails to explain the elements.

Hegel's, Phenomenology moves from "Sense-Certainty" to "Perception" and then to "Understanding." In Understanding the "Object is now its Concept." Hegel contends that our desire is "Freedom" or autonomy. Kant distinguishes "understanding" and "reason." Understanding applies concepts to experience. In contrast, reason reflects on concepts, logic, metaphysics and theology. Hegel accepts this difference. Reason synthesizes and analyzes objects and events.[23]

71

### Logic, Dialectic and Physics

Hegel argues "reason" is "purposive activity." He does not think that all things are in the mind; instead, he argues there is a unity of consciousness and nature. Both are rational. The "It" in Physics "is us." There is unity between Subject and object. The absolute is similar to a principle of complexity rather than a final solution. It is an irreducible principle.[24]

Concepts are not just attributes of consciousness but features in nature. We see gravity, for example, in the falling of a stone. The end or teleology of nature is the "universal individual--the living Earth." The brain is indistinguishable from consciousness or the "Spirit." Knowledge is a unified whole with an end--the Spirit. We are not merely machines. Descartes unites thought and Being. Hegel thinks that self is a <u>process</u>.[25]

The theme of the Master-Slave parable is that self-consciousness is complex and dependent on other people. Action like a fight to the death "contains a double significance." There is a reflection back and forth awareness. Self consciousness is a relationship of Master and Slave and is a product of a "life-and-death" fight. The Master becomes dependent on the slave; the slave becomes independent of the Master."[26] Slaves were property. A completely materialist explanation of the universe is slavery.

Logic, Dialectic and Physics

Henry A. Erikson's <u>Retrospect</u>

Theses: We are also the "It" in Physics.
Physics is a Group Guess
Hegel is Existential: "Existence is Logic" Mike Franey
Subject and Object are United at Minnesota
"Growth, Metamorphosis," or Dialectics are the Keys to the
History of the Physics Department

Erikson presents what the writer calls the historical **conversation** or **journey** of the Physics Department at the University of Minnesota. He looks at the continuity, the transition and the development of the department. The thesis that the writer draws from Erikson's research is that the microcosm of this history and the macrocosm of the universe unite. For example, the department produced Ernest O. Lawrence who built the cyclotron.

Erikson writes that John Zeleny wrote the history of the physics department from the beginning to 1914-1915. Henry A. Erikson wrote the history of the department to 1938--a total of 23 years. This history includes his years as chairperson of the department. This document, <u>Retrospect</u>, Erikson wrote during 1938-1941 after he retired. The manuscript contains many statistics. The research contributions of the student and faculty are their publications.[1]

Erikson became a member of the physics staff during the 1897-1898 school year. Before, he taught science for one year in the Rochester, Minnesota, High School. Erikson received his Bachelor's of Electrical Engineering in 1896. The Mayo Brothers started the Mayo Clinic during these years. The high school movement was growing and some small colleges were closing. In these years there were about 400 Minnesota High Schools. The State Board set the standards for these high schools. To get into college a student had to pass high school examinations.[2]

### Physics in Nineteenth Century America: A Background

What is significant about science in nineteenth century America? Nathan Reingold suggests that this century was an important century in the history of science, but American contributions did not predominate. He contends that the reason for this is that Americans emphasized applied research instead of basic or theoretical research.[3]

73

Logic, Dialectic and Physics

## Geography and Physics Allied

Joseph Henry, a physicist, did not found the American school of physicists. Benjamin Pierce, the mathematician, did not break the turf in pure mathematics. However, Wolcott Gibbs, the Harvard chemist, developed a group of research chemists. The field of astronomy had a high reputation in America. However, it was geography that led the sciences in America for most of the nineteenth century. This field had two parts, geophysics and natural history. Geophysics emphasized the physical properties of the planet, specifically the dimensions of the earth's surface, its atmosphere, the oceans, gravity, meteorology and the earth's magnetism. In contrast, natural history dealt with the nonexperimental elements of biology and geology with an emphasis on classification. These two fields merged in geography where developmental surveys and nature expeditions coalesced.[4]

The main difference between these fields was that geophysics used mathematics a great deal while natural history did not. Most physical scientists did not focus on physics, mathematics or chemistry. For example, chemists analyzed rocks and minerals collected in the field. Physicists studied the earth's magnetic field, astronomers studied the clouds and winds, and mathematicians defined the size and shape of the earth for data. Four fields received the most contributions from American research: astronomy, botany, geology, and meteorology. Each of these fields came together in geography.[5]

## Joseph Henry: Physicist

Joseph Henry advocated the field of geography while at Princeton and as the Secretary of the Smithsonian Institute during the period 1846-1878. As a physicist he studied electricity and magnetism. He built electromagnets. Henry was a leader in the American Philosophical Society and in the Franklin Institute both located in Philadelphia.[6]

## The Geophysical Group II: Henry, Bache and Others

A single group of scientists called the "Lazzaroni" meaning scientific beggars dominated the American scientific scene before the Civil War. The guiding spirits of this group were Joseph Henry, Alexander Bache and Benjamin Pierce, a mathematician. They focused their efforts on geophysics. Louis Agassiz, the naturalist, Wolcott Gibbs, the chemist, and also Gray, the botanist, were the exceptions to this group. Matthew Maury, an oceanographer was the rival.[7]

Logic, Dialectic and Physics

After Joseph Henry began in the Smithsonian Institute in 1846 he coordinated a system of telegraphic weather records. By 1860 nearly five hundred weather stations sent meteorological reports to the Smithsonian Institute. Joseph Henry by 1850 displayed daily weather maps and reports. Today's Weather Bureau developed from the Smithsonian system.[8]

### The Birth of the National Academy of Sciences

During 1863 Congress passed a bill signed by Lincoln starting the National Academy of Sciences. Senator Henry Wilson of Massachusetts sponsored the bill. In section one the legislation was straightforward. It named fifty Americans as founding members. Section two set the membership at fifty and gave the Academy the right to elect participants and to make rules. Section three gave the academy the right to report questions of "science and art."[9]

The forces behind this legislation were Alexander Bache and the Lazzaroni or "scientific beggars." It was an elitist group compared to the American Association for the Advancement of Science which was open to all supporters of science. The Lazzaroni asked for funds from the government for scientific research. European Academies influenced this group. The Civil War made many scientists willing to serve the Union.[10]

When the problem of helping the Union arose, Joseph Henry opposed the formation of an Academy. It would politicize the scientific community. Henry, Bache and Charles Davis started to promote a committee to counsel the government on patents and inventions. A permanent commission in the Navy Department was the end of this effort, and it began on February 11, 1863. It advised the Congress and Lincoln for the rest of the Civil War.[11]

The Lazzaroni influenced the composition of the National Academy. Thirty-two incorporators were physical scientists and natural historians. It did not include three noteworthy scientists. The first was George P. Bond, the head of the Harvard Observatory, who had antagonized Pierce, Bache and Gould. The second was Spencer F. Baird, a prominent naturalist and the assistant Secretary at the Smithsonian. Agassiz opposed him. The third was John W. Draper, a chemist with an international reputation, who they did not select because he was too unorthodox.[12]

At the opening meeting Bache became the President of the Academy. His main opposition was William Rogers (1804-1882), a geologist who started the Massachusetts Institute of Technology. By the next meeting in 1864

### Logic, Dialectic and Physics

Bache, the leading Lazzaroni, had a stroke and Baird became president. He willed his estate to the new Academy, and saved the Academy. After Bache's death, the group transformed itself into a learned group with branches. It was not until WWI that the National Academy built an operating movement, the National Research Council. It then became a strong force for scientists.[13]

Between the Civil War and WWI some organizations for specific scientific research appeared. Reingold argues that the scientific community spurned power and tended its own gardens. There were not a lot of funds for scientific research during this period. The Lazzaronis lost out to Henry and his cohorts.[14]

### A Revival of Physics

People did not think that physics would be a field making major scientific contributions in twentieth century America. Looking backward the major American names--Rowland, Michelson, and Gibbs--in physics surpass the researchers in the biological sciences. Significant also are John Draper (1811-1882) and Henry Draper (1837-1882). They helped professionalize physics.[15]

The Drapers taught at the Medical School of New York University. John Draper made a contribution to photochemistry. He was one of the first to photograph a human face. John Draper also photographed the moon. He analyzed the radiation and photographed the solar spectrum. Henry, his son, continued to do research on the solar spectrum. His major success was to photograph a star which showed Fraunhofer lines. Henry Draper's research provided a catalog of stellar spectra by Harvard Observatory. Contemporaries consider the Drapers amateurs because they did not got to graduate school in physics. There were physicians gone awry.[16]

### Henry Rowland

During 1871 a young researcher, Henry Rowland, at Rensselaer Polytechnic Institute at Troy, New York submitted a paper to Silliman's Journal. His paper was on magnetism. The journal first accepted and then rejected Henry Rowland's paper. His second paper, "On Magnetic Permeability and the Maximum of Magnetism of Iron, Steel, and Nickel" he submitted to England's great physicist, James Clerk Maxwell (1837-1879). Maxwell published the article.[17]

Logic, Dialectic and Physics

After this success, Yale scientists accepted him to a Ph.d program. When Daniel Gilman (1831-1908) heard about Rowland, he offered him a position at Johns Hopkins University in Baltimore. Rowland agreed and went overseas for his education. When a colleague of Rowland's at Hopkins, Ira Remsen (1846-1927), a chemist, discovered that Silliman's Journal would not produce his students, he started a new journal, the American Chemical Journal during 1879.[18]

Rowland had a professional career in experimental physics. For example, he analyzed light using a diffraction grating. Rowland invented gratings that had nearly one hundred thousand lines six inches long. This grating was the most refined instrument in the field until Michelson's echelon grating surpassed it. The power of resolution related to the number of lines and these lines must have the same shape and be parallel. Also, Rowland did research in electrical engineering. He called for a pure science.[19]

### The Speed of Light: Newcomb and Michelson

During 1878 Albert A. Michelson (1852-1931) was an ensign in Annapolis in the Department of Physics and Chemistry. Simon Newcomb (1835-1909) was the Superintendent of the Navy's Nautical Almanac office. He was a mathematician. Both of these men had humble origins.[20]

Newcomb was a Canadian who had an unhappy youth and ran away to America. Joseph Henry recognized his ability and got him a position in Cambridge. He got to attend the Lawrence Scientific School at Harvard. Newcomb was a classical mathematician and astronomer. In contrast, Michelson was a researcher in experimental optics. Michelson, at an early age, came with his family to America from German Poland. Annapolis accepted him.[21]

Newcomb wanted to determine solar parallax. His greatest efforts arose with his association with Michelson. In 1878-1785 they tried to test solar parallax by studying the transit of Venus and by analyzing light's velocity. By 1867 Newcomb tried to determine the velocity of light as a method for researching an accurate measure of the radius of the earth's orbit. Not known to him Michelson had found a modification of Foucault's revolving mirror approach in the fall of 1877.[22]

In March of 1878 Newcomb came across Michelson's work. Michelson's experiment was a great discovery. He got results in January 1879. Following this, Michelson went to the Nautical Almanac to help

Newcomb with his experiments. In 1880 Michelson went to Europe to study for two years. He invented an interferometer for his research. Alexander Graham Bell helped him financially.[23]

The results of Michelson's and Morley's research in Cleveland during 1881 brought about doubts on several basic premises of traditional physics. At first Newcomb's and Michelson's results were as of 1879 200 km/sec. different. Newcomb did not get money from the Bache Fund to further determine the velocity of light. Eventually at the Case Institute, Newcomb and Michelson got similar results.[24]

Reingold depicts Michelson's and Newcomb's relationship to that of a graduate student and his advisor. For example, Newcomb gave Michelson the Works of Foucault, his predecessor. Michelson passed up Newcomb, Reingold contends. [25]

### Josiah Gibbs (1839-1903)

Josiah Gibbs was the unique product of several research fields in America. His father was a language professor at Yale. Now at Yale there was a budding science and classics tradition. Silliman started the science tradition and James Dana continued it. These two thinkers established in 1846 the United States graduate school and the Department of Philosophy and the Arts. In 1854 an undergraduate science school called the Sheffield Scientific School started.[26]

Gibbs did his doctoral dissertation "On the forms of Teeth of Wheels in Spur Gearing." Later, he built a brake for trains and invented a governor for steam engines. During 1871 Gibbs became a Professor of Mathematical Physics in the Department of Philosophy and Arts. He wrote several theoretical works. James C. Maxwell was one of the first to recognize Gibbs findings. Gibbs was then an exception to the theme that Americans cannot do theoretical work in the sciences. He did work in thermodynamics and dynamics and developed vector analysis and laid the groundwork for physical chemistry. Specialists recognized his achievements in these fields.[27]

### The Goals of a Physicist by Henry Rowland

At the time Rowland began to study the physics, a physics organization did not exist. He was the first President of the American Physics Society. In his valedictory to the sciences he contended that curiosity was an acceptable end in itself.[28]

Logic, Dialectic and Physics

Rowland said, we form "an aristocracy of intellect." Later, he argued that absolute truth or absolute falsehood do not exist. We should calculate the probabilities of research. For example, Faraday's and Maxwell's research showed the world that the idea of electricity as a fluid in nature is in error.[29]

## New Vistas in Physics: Minnesota

The years 1895-1897 were a new age for physics. For example, scientists discovered x-rays, radiation, and the electron. What were some of the discoveries? Erikson suggests scientists used the "Atwood machine [and] Oersted's piezometer in demonstrations. Also, they projected thallium, bismuth and strontium lines by evaporating the metals in an arc. Scientists also projected the reversal of the sodium line by a sodium gas flame."[30]

What were some of the other experiments? There was the freezing of water through evaporation in a vacuum, the galvanic skin response, the deflection of a needle by a current, electrolysis, Hertzian waves, and Faraday's induction experiment. Furthermore, after Rontgen discovered x-rays, a person from Wisconsin came to get a bullet in his body located by x-rays. The doctors located the bullet. Mentioned earlier, it was J.J. Thomson who discovered the electron. Becquerel did work with radioactivity. Marconi studied Hertzian waves and invented the wireless.[31]

Few physicists were aware of the great scientific revolution that was starting. Graduate schools began to grow. However, from 1892 to 1896 Erikson studied by a kerosene lamp. Gas lights lit the class rooms. Professor Jones came to class at the University of Minnesota in a horse-drawn carriage. Students and faculty could not smoke on campus. The university did not encourage outside work.[32]

What were some of the changes that occurred during Erikson's life? He remembers the world moved from horses and wagons to automobiles and airplanes, from steam engines to combustion engines, from kerosene lanterns to fluorescent lights, from the telegraph and radio and the ice person to the refrigerator, from the hand plow to the combine, from the theory of spontaneous generation to the germ theory, from unsanitary conditions to indoor plumbing, from the stage to motion pictures, from atomic theory to intra-atomic research, from classical physics to relativity and quantum mechanics. Physics laboratories became obsolete over night. Measurement became much more precise. For example, in 1889 the physics department purchased a Geneva Society cathetometer and a standard meter. Erikson remembers

79

## Logic, Dialectic and Physics

hearing Professor A.A. Michelson saying: "They cannot destroy our beloved ether." During 1900 Planck's quantum idea appeared. A mental avalanche followed. The foundation, Erikson contends, of classical physics, specifically, the "law of cause and effect hung in doubt."[33]

### Back in Minnesota

As of 1897 the physics staff at the University of Minnesota contained four members: Professor F.S. Jones, the department head, John Zeleny, Anthony Zeleny and Henry Erikson. There were no graduate courses offered. There was a course in mechanics and one in light offered. The students created projects. Anthony Zeleny got his M.A. degree in 1893. His thesis was on Hertzian Waves. John Zeleny received the first doctorate in 1906 and Anthony Zeleny got the second in 1907.[34]

### Physics and Its Germination

In 1900 few could have contended that physics was a great discipline. Today, echoing Newton, physics is an envied field. It is a model field. There are several noteworthy physicists from this era. Albert Michelson, for example, was the first American to receive the Nobel Prize (1907). However, there was still a feeling of inferiority among the American physicists. [35]

The cutting edge of physics was still in Europe. European physicists came to American Universities to give lectures. A second problem of physicists was poor pay. In Europe, physicists received more economic and intellectual recognition. Physics changed its status first by Faraday and later by Einstein. The rise of the status of physics in twentieth century America reinforced the research efforts of all scientists.[36]

### In Minnesota

In 1895 Dr. Henry J. Eddy began teaching engineering mathematics and mechanics. In the twentieth century he became the first graduate school dean at the University of Minnesota. Students analyzed the thoughts of physicists like, Maxwell, Abraham, Drude and A.S. Webster in depth. Scientists taught special topics like the Stark and Sieman's effects and Lorentz's ideas. Professor Frederick Sheets Jones, the first chair of the physics department, studied in Germany under Helmholtz, Kundt and Wehr. Jones introduced Wehr's galvanometer, telescopes and scales. These items became obsolete when the suspended coil and permanent magnets came into use. Leeds and

Logic, Dialectic and Physics

Northrup designed these items. Professor Jones became the dean of the Engineering College.[37] Hegel argues that ideas magnify sensory data.

## The Microcosm is the Macrocosm

From 1890 to 1900 the physics department was in the east side of the chemistry building. The department shared quarters with the Electrical Engineering department. In 1900 the physics department moved into the armory. In 1899-1901 the physics department received $75,000 to have a new physics building.[38]

During March 1, 1897, John Zeleny went on leave to Germany and England where he studied under J.J. Thomson, Rutherford, Townsend and other scientists. Zeleny studied the ionization of gases at the Cavendish Laboratory. During October 1902 the University of Minnesota Physics department moved into Jones Hall, the new laboratory. A Hampson air liquefier and a White compressor became staple tools. Low temperature research was in vogue. Also, the department got a shop with milling machines and lathes. Jones Hall was the department's home from 1902 to 1927 when it moved into a new laboratory on the university's mall. It is there today.[39]

Erikson suggests that Jones Hall, the new laboratory, was like a new universe. Scientists discovered the electron. Also, Roentgen ionized gases. Rutherford had begun working with radioactivity. Wilson saw that ions could act as "condensation nuclei." The University of Minnesota Graduate School promoted work in physics. In 1902 Alois F. Kovarick became a student. Henry Erikson received his doctorate in 1908 and A.J. Kovarick in 1909. Professor Jones became a dean at Yale in 1909.[40]

## Back on the National Scene: The National Academy's Reform

From the beginning of George E. Hale's election in 1902 to the National Academy of Sciences, he worked to revitalize the sciences and arts. This academy began in 1863 but it did not fulfill the aspirations of its members.[41] Its status was tentative.

What was the strength of this organization? It embodied the traditions of science. Hale who did research on the sun, emphasized internationalism for the academy. Hale wanted to create a scientific culture to deal with the complexities of knowledge and civilization. Evolution was the key to this culture.[42]

## Logic, Dialectic and Physics

Hale's ideas for reform tried to meet the needs of specialized educational societies. He tried to bring them under the control of the academy. Hale wanted to expand the National Academy to include many engineers, social scientists, and the humanities. Many of the members lacked enthusiasm for Hale's reforms. To lead the scientific community Hale promoted the academy.[43]

While Hale worked to expand the academy, James M. Cattell developed the American Association for the Advancement of Science to be the ruling body of learning in America. In contrast to Hale, Cattell used a federal system of government to represent the variety of learning societies. Cattell desired to work with the political system. In contrast, Hale mistrusted the political process.[44]

Reingold suggests that the diffuse pluralistic American society frustrated Hale's reforms. Many institutions blocked Hale's vision. American had a science culture that emphasized research. Even today many of the goals of scientists do not get adequate recognition thinks Reingold.[45]

### The Minnesota Connection

During 1908-1909 Erikson did research at the Cavendish Laboratory in Cambridge, England on how ions recombine at various temperatures. Professor Thompson became a knight during that year. Professor C.T.R. Wilson got his expansion chamber "tracks" or marks. Professor Rutherford received the Nobel Prize.[46]

Back in Minnesota Cyrus Northrup of Yale became President of the University in 1885. He resigned at age 68 at the end of the 1910-1911 school year. George E. Vincent of Chicago became President on April 1, 1911. He remained President until 1916-1917 when he became President of the Rockefeller Foundation. The social science building at the University of Minnesota bears his name. During these years Dr. Guy S. Ford became head of the Graduate School. Dr. L.D. Coffman headed the School of Education; and, Dr. E. Lyon was leader of the Medical School. Professor John Zeleny became the head of the Physics Department in 1909-1910.[47]

Dr. Kovaric worked with Dr. Geiger and Niels Bohr in Manchester. He became an associate professor in 1915. Dr. J.J. Thomson visited the laboratory in 1909. While Professor John Zeleny was on a sabbatical in 1914-1915, Professor Anthony Zeleny became the acting head of the physics department. In 1914-1915 Dr. John Zeleny accepted a position as head of the

Sheffield Scientific School at Yale. Crystal radio sets were a topic of interest. The social sciences, business, economics and sociology grew up during these years.[48]

Dr. Erikson became chairperson of the department in 1915-1916. He remained in the physics department until he retired in 1938. WWI was in progress. Edison phonograph records played at academic tea gatherings. Students did research on ionization potentials and atomic theory.[49]

Dr. Compton found the "Compton effect" where a photon has particle properties.[50] He got a Nobel Prize for his efforts. Compton got a position at the University of Chicago. There, he studied the scattering patterns of magnets on x-rays. He used ionization spectrometers to record his results. John Zeleny analyzed "ionization energies in hydrocarbons."[51]

After the sinking of the Lusitania in 1917 America became involved in WWI. Dr. John Tate was one of several professors who left the physics department to work in the war effort. Also, in 1917-1918 Graduate School Fellows became available. As early as 1916 a movement for a new laboratory emerged. This new laboratory effort culminated in the creation of the present physics laboratory beginning in 1927.[52]

## World War I: Progress?

Reingold suggests that progress impaired that outlook of many people during the late nineteenth and early twentieth centuries. For example, industrialization spread, illiteracy was almost eradicated, and many horrible diseases wiped out. Science fueled progress.[53]

Not all accepted progress. Some felt that we were regressing. However, steam powered transportation, the telegraph, and the press speeded up the research and exchange of ideas. Science became universalized.[54]

Was World War I progress? Maybe it was not! People felt that civilization was falling apart. The war horrified Americans. British propaganda and German arrogance brought the United States into the conflict. Few American scientists favored the Germans. However, Americans emulated German research.[55]

In response, Jacques Loeb, for example, opposed the martial temperament. As a Jew, he was fearful about how nationalism could turn into racism. After the war, he questioned the civilizing aspects of science. Other scientists questioned the idea of the internationalism of science. Atrocities in Belgium, the slaughter at Verdun, and submarine warfare cracked the facade of science. The world ostracized Germany. They barred Germany from the League of Nations until 1926. WWI changed the old order.[56]

George Hale favored the allies. He worked for preparedness. Hale had some rivals: the Naval Consulting Board, and the Committee of 100 of the American Association for the Advancement of Science. This committee did not try to coordinate science but to enhance research. Hale backed the National Academy. He was a conservative. The national position called for defense readiness.[57]

By WWI, the Rockefeller Foundation expanded its operation. By 1915 the Committee of 100 asked the Foundation about backing for scientific research. The Naval Consulting Board did not challenge Hale's program. Thomas Edison headed this Board. The efforts of these scientists and engineers led to the beginning of the Naval Research Laboratory. The academy created a National Research Council. This Council by 1917 found a place in the armed services. Robert A. Millikan administered its program. A presidential order kept the council going at the end of the war.[58]

During 1918 the Rockefeller Foundation raised the issue of creating a research body for chemistry and physics. Millikan favored placing research bodies at universities. Hale thought, for a while, that the institute could exist under the control of the academy. A compromise developed--the National Research Council Fellowships during 1919 began. Many favored leaving science in the private sector. Defense was an exception to this program. After the war, the National Research Council and the fellowship program, Hale's projects, survived. Reingold argues that during 1920-1940 scientific culture grew rapidly. The National Research Council could not contain the diversity of the scientific community.[59]

What were the problems of science between the World Wars? Reingold argues first that status in society troubled them. Second, a mass scientific community arose. Could this community of scientists enhance democracy?[60]

### Physical Sciences between WWI and WWII

Reingold suggests that the more complicated and abstract a topic the harder it is to get funding. This was the case of the physical sciences between the world wars. Chemists and physicists felt that both theory and experimentation were essential. Astronomy was important. Mt. Wilson and

## Logic, Dialectic and Physics

Mt. Palomar came into existence. Einstein displaced "classical physics" with "modern physics." Physics became the benchmark or model for the other disciplines.[61]

After WWI, industrial research grew rapidly. For example, General Electric and Bell Telephone laboratories produced two Nobel Prize winners: Langmuir and Davisson. There was a dramatic increase in the role of physics. The National Research Council Fellowships backed postgraduate research. Physics was an evolutionary field.[62]

Two of the most important mathematicians were Norbert Wiener and John Von Neumann who expanded theoretical physics. The field of mathematics became like hard currency during these years. Oswald Veblen built the Princeton School of Mathematics by forming the Institute for Advanced Study.[63]

Felix Klein (1849-1925), a mathematician argued that physics was too vital to leave it to the physicists. On another front Ernest O. Lawrence built up high-energy physics. J. Robert Oppenheimer and Wiener magnified the relationship between mathematics' theory and physics' practice.[64]

### Princeton's Institute for Advanced Research

Abraham Flexner thought up the idea for the creation of the Institute for Advanced Study. The money for the institute came from Louis Bamberger and Mrs. Felix Field. It started in 1930 and began operating at Princeton, New Jersey in I933. Flexner was the first head of the institute. It was one of the first think tanks. Oswald Veblen was the chief of Princeton's Institute's School of Mathematics. Flexner wanted the institute to go beyond graduate school research. It would be America's zenith in higher education. Flexner wanted full-time research.[65]

Hitler clashed with Flexner and Veblen. These two scholars tried to help scientists and scholars flee Germany. Einstein came
to the institute. He spoke out for victims of Fascism.[66]

Veblen wanted to make economics a "clinical" field. Flexner favored lay control and isolation from the world for this institute. Flexner was a conservative and he opposed the New Deal. He wished to avoid controversial areas of research. When the threat of a German atomic weapon developed, Einstein signed a letter that brought about the Manhattan project. Hiroshima and Nagasaki followed.  These events shattered the idea of a science separated from human history.[67]

## Minnesota

Back in Minnesota the machines used for research were becoming larger. For example, by 1939 the department built the Van de Graff generator outside the physics building.  Moreover, a 20 by 1 million volt cyclotron would be as large as one third of area of the physics laboratory at Minnesota.  The physics research, now, had remodeled the elementary atomic structure.  Particle wave duality emerged.[68]

Erikson contends that physics is the "mother science."  It spun off electrical engineering, radio research, meteorology, bio and physical chemistry, radiology and industrial laboratories.  To meet the needs of military training, the University Senate changed from the semester system to the quarter system on December 19, 1918.  The department put a professor at the head of each division:  "mechanics, heat, electricity, optics and acoustics."  Professors preoccupied themselves with three lectures per week and teaching assistants corrected the quizzes.  Class and lecture sizes increased.  Professor L.F. Miller taught heat.  A. Zeleny taught electricity.  Dr. J. Valasek taught optics.  And, Dr. J.W. Buchta taught acoustics.  Erikson taught mechanics.[69]

Erikson argues that the greatest educational problem of the time was "can a student be as efficiently taught in a large group as in a small group?" The College of Education  and the Physics Department studied this problem and found  "there was no perceptible difference."  If there was a variation, it depended on the subject taught.  After President Woodrow Wilson declared war on Germany on April 6, 1917, many of the professors left for war research.[70]

In 1918-1919 Dr. W.F.G. Swann came to the physics faculty and he did research on, among other items, cosmic rays.  Later, he went to the University of Chicago, Yale, and the Bartol Research Laboratories at Swarthmore.  Dr. L.F. Miller taught at both the University of Minnesota and the School of Mines in Golden, Colorado.  When he came back to Minnesota, Miller taught heat. During 1920 the Physics Department gave Miss Vina Downey the first Ph.d given to a woman.  She did research on textiles.[71]

WWI ended for America on November 11, 1918.  On January 1, 1919, Dr. John Tate came back to the physics department.  Dr. McKeehan went to Yale, and Dr. Klopstig went to work at the Research Laboratory at the Leeds and Northrup Co.[72]

Logic, Dialectic and Physics

In 1928-1929 Professor Tate began teaching Theoretical Physics which became a prerequisite course to graduate study. Erikson suggests that Tate's name became "synonymous" with that course. In 1938-1939 Dr. J.W. Buchta succeeded Erikson as the head of the physics department.[73]

In 1921-1922 Dr. John G. Frayne joined the staff. He was from Ireland. Frayne had before gone to Ripon College. At the end of the teaching year he took a position at Antiock College, Ohio. Also, Merle Tuve and Ernest O. Lawrence enrolled as graduate students in the department. Tuve did research on terrestrial magnetism. Lawrence joined the University of California where he built a cyclotron and got a Nobel Prize for his efforts.[74]

In 1921 John Zeleny, doing research at the Cavendish Laboratory, found "the mobility in air of the negative ion was greater than that of the positive ion in air by a ratio of about 1.4." This difference in mobility remained unsolved for twenty-five years. Erikson developed a method where he forced ions into a plate having an electrode connected to an electrometer: A blast method where ions entered an air stream by an electric field where the researchers measured the electric current at different points down the air stream. They found a curve that measured ionic mobility. Erikson measured the curved for positive and negative ions. The result was a maxima that showed equal mobility. When Erikson used "aged" ions, then the positive maximum was stationary given the mobility ratio of 1.4. Thus, the "initial" positive ion had the same mobility as the negative ion. This ion did not have time to attach itself to another molecule.[75]

The physics staff lost a key member in 1922-1923--Dr. Swan. He went to the University of Chicago. The department hired two younger men to replace him. They were Dr. John H. Van Vleck of the University of Wisconsin and Dr. Gregory Brect of Washington D.C.'s Carnegie Institute. Dr. Brect worked with the Department of Terrestrial Magnetism and returned to the University of Wisconsin to be a professor of Theoretical Physics. Dr. Van Vleck, after a short stay at Minnesota, went to the University of Wisconsin and later to Harvard. Also, Dr. E. Condon went back to Princeton and then joined the staff at Westinghouse Research Laboratory. Salaries rose dramatically during the 1920s.[76]

In 1930-1931 Dr. J. Frenkel of Leningrad's Physical Institute was a guest professor. He wrote part of a book on "wave mechanics" while he was here. Frenkel was not a communist party member. He returned to Leningrad. In 1925-1926 Edward L. Hill started in the department. After receiving a doctorate, he devoted his life towards studying theoretical physics. In 1934-

**Logic, Dialectic and Physics**

1935 Dr. John H. Williams joined the department. The department named Williams Laboratory after him.[77]

 Erikson contends from 1900 to 1938 the physics department was host to many famous physicists. For example, Sir J.J. Thomson, his wife and son, visited the department. Professor Shrodinger spoke on new wave equations. Professor A. Sommerfeld gave a lecture on "fine" structures. Professor P. Dirac lectured on quantum mechanics. Sir Oliver Lodge gave a lecture on "communication with the Spirit world."[78]

 During 1937-1938 Dr. Erikson and Dr. Anthony Zeleny retired. The department gave a dinner for them on May 24, 1938 at the Curtis Hotel. A small galvanometer was given as a present to Dr. Zeleny, and a camera was given to Dr. Erikson. In retrospect, Erikson concludes that he had no wish that his "lot in life" would have been different.[79] A theme that highlights the physics department is the principle of the man of science.

Logic, Dialectic and Physics

Nier "My God to Thee:"  Retrospect

University of Minnesota Physics Department Chairmen

| | |
|---|---|
| Frederick S. Jones | 1889-1909 |
| John Zeleny | 1909-1915 |
| Henry Erikson | 1915-1938 |
| J. William Buchta | 1938-1953 |
| Alfred Nier | 1953-1956 |

During 1940 there were eight faculty in the department. Between 1920 and 1930 J.H. Van Vleck taught quantum mechanics. Research was done on the electron impact on gases. Joseph Valasek analyzed x-rays. Scientists discovered ferro-electricity or electric fields in crystals. John H. Williams built up nuclear research between 1920-1930s. Shrodinger and Dirac visited the department. Williams went to Los Alamos. In 1938 John Bardeen studied the theory of low temperature phenomena. Later he helped invent the transistor and the super conductor. No research was done during the WWII years.[1]

In 1945 the department came "back to life again." Cosmic ray research using high altitude balloons began. Frank Oppenheimer continued his career at Minnesota. The cosmic ray program continued between 1950-1980. The Van De Graaf Generator became obsolete at the end of the war. The linear accelerator built in the 1950s became obsolete in the 1960s. John Williams continued nuclear research during these years and the department named the nuclear laboratory after him. During the 1960s research was done on condensed matter, on solid state physics, and on low temperature physics. During the years 1950 to 1970s Nier crystallized the field of mass spectroscopy. Research on electrons continued. The most significant date in the history of the physics department at the University of Minnesota was July 1, 1959, when the physics department became a school in the Institute of Technology.[2]

## Two of Alfred Nier's Articles

Alfred O. Nier who was a long time member of the University of Minnesota's Physics Department made a significant contribution to the Manhattan Project. He helped develop and perfect mass spectrometry which is a tool used to study dissociation and ionization of molecules after electron impact. Mass spectrometry determines the abundances of isotopes.[3]

## Logic, Dialectic and Physics

What stimulated this research? The petroleum industry needed a method to analyze complex hydrocarbon mixtures, and the atomic bomb program, the Manhattan Project, needed to determine uranium isotope separation. Nier moved this tool from use by specialists to mass use and production. Today scientists use this instrument to analyze gases, to follow isotope tracers and to determine isotope abundance in geology and the cosmos.[4]

After completing his studies at Minnesota in 1936, Nier got a National Research Council Fellowship. He used it to study at Harvard with Kenneth Bainbridge. In his thesis Nier measured the frequency of isotopes in several elements--"argon, potassium, zinc, rubidium and cadmium." Bainbridge suggested that Nier measure the heavy elements such as lead and uranium. Together they designed a mass spectrometer that could measure isotopic abundance ratios for the full atomic table. This analysis needed an electromagnet that weighed two tons and a 5 kw generator with a stabilized output voltage to run it.[5]

In the 1930s Cambridge, Massachusetts was a bed of scientific culture. For example, Alfred Lane was measuring geological time. Robley Evans, an MIT professor, was analyzing the radioactivity of minerals. At Harvard Gregory Baxter was making atomic weight measurements on lead extracts from uranium and thorium.[6]

### Uranium Isotopic Composition

During the 1930s the "isotopic abundance ratio, 238 U to 235 U," was not precisely known. Working with Baxter, Nier measured the ratio, 238 U to 235 U at 139. Later, measurements found it to be 137.8. Also, they discovered the isotope, 234 U occurring in 1 part in 17,000 in uranium.[7]

After returning to Minnesota in 1938, Nier worked with John Tate to get funds to obtain a 2-ton magnet like the one at Harvard. He used the mass spectrometer tubes that he had designed and built at Harvard. Nier redesigned the mass spectrometer and reduced its size to several hundred pounds. This instrument became a prototype of those used for the Manhattan Project. Physicists called it "the Sector Mass Spectrometer" because it had a 60 degree sector magnet on it. This magnet was run by automobile storage batteries.[8]

Scientists discovered nuclear fission a few months before the American Physical Society Conference convened in Washington, D.C. during April 1939. 235 U caused the slow fission of uranium. Nier thought that he could use his 180 degree mass spectrometer to get enough 235 U to have a test.[9]

Logic, Dialectic and Physics

To complete this separation Nier used UF6. This compound was volatile and coated much of the surface of the mass spectrometer tube. He created a new mass spectrometer tube, and formed an oven into an ion source and ionized the vapor with an electron beam. Nier found there was no contamination in the background. Scientists conducted tests during February 28-29th., 1940. The scientists gathered two samples of separated 238 U and with several Columbia University researchers they hit the targets with neutrons. The result was that 235 U produced the fission fragments. A great step forward in science had happened.[10]

A Columbia group had a different approach. They wanted to separate 235 U by gaseous diffusion. Another group including Jesse Beams at the University of Virginia and Phil Abelson at the Naval Research Laboratory tried a third approach. They tried liquid thermal diffusion. These two groups did not have a spectrometer to check the performance of their systems. Only Nier had a mass spectrometer that could check "uranium isotope abundance ratios."[11]

Nier got funds from the Office of Scientific Research and Development to perfect uranium analyses. Scientists built out of these efforts a K-25 separation plant at Oak Ridge, Tennessee. In 1942 Columbia and Virginia, including Mark Inghram and Ed Ney, constructed seven instruments to analyze uranium isotopes. This program came to be known as the Manhattan Project. These scientists built in 1943 12 instruments for hydrogen analyses in a few heavy water plants that produced deuterium. A decision germinated to start a gas diffusion plant. The plant had to be a vacuum because it would function below atmospheric pressure. Also, it would use UF6 gas. This gas reacts with water thus it did not clog the diffuse membranes. Nier talked with John Dunning and Eugene Booth at Columbia about using a mass spectrometer to find leaks.[12]

Nier decided that he would build a portable mass spectrometer. His group substituted the glass spectrometer tube and pump with metal. Later, Nier moved to New York to lead an instrument laboratory at the Kellex Corporation. There, he constructed an "on-line system" for finding the impurities in the "process stream" in many locations. General Electric contracted to build 100 instruments containing a strip chart recorder that contained a slave recorder. One person could check a large plant. Thus a person could monitor impurities .[13]

**Memories**

## Logic, Dialectic and Physics

Nier worked with individuals studying the bombardment of atoms and molecules by electrons. After doing takes, they could check the abundance of various isotopes. Thus, civilian scientists played an important role in military research. This effort led to the building of the Office of Naval Research. It led to the government support of programs like the National Science Foundation. A highlight of Nier's career was a letter from Enrico Fermi encouraging Nier's research.[14]

## A Second Article by Al Nier

Bleakney invented an electron impact ionizing system during 1929. This system played an important part in mass spectrometry research in the 1930s. Scientists made an instrument that replaced the solenoid with an electromagnet.[15]

## Background

J.J Thomson constructed the first mass spectroscope. He created ions in electrical discharges. Thomson put them through electric and magnetic fields which created beams that had a correspondence of mass to ionized charge. Then, the ions hit a photographic plate. The placement of the darkened traces on the plate relates to the masses. The intensity of the darkening relates to the abundance. Aston built an instrument that had an improved array of fields. it compared to an "optical spectrograph." Dempter built the first mass spectrometer. In this device he measured ion currents by electrometers. Thus, he could measure abundances of ions.[16]

## Bleakney's Device

In the 1920s and 1930s John T. Tate was studying the idea of electron energy levels in atoms. So, he studied electron impacts. Bleakney was a graduate student of Tate's. Nier studied with Bleakney who was his lab instructor. Bleakney's mass spectrometer was in a glass tube and held within a solenoid. E.U. Condon, a physics faculty member at the University of Minnesota, prompted Bleakney to study ions created when H2 ionized. Bleakney found energized H + ions and thus verified the Franck-Condon effect. Later, Bleakney showed when he used a "retarding potential" his ions had kinetic energy. He went to Princeton and built a mass spectrometer with a 180 degree mass analyzer.[17]

92

## Logic, Dialectic and Physics

Nier became a physics thesis student during 1934. There was then a big innovation in the production of electrometer vacuum tubes. These tubes improved the creation of mass spectrometers. Nier used this technology to study complex molecules. Working under Professor Tate, Nier used this instrument to study nuclear physics.[18]

Along with Professor K.T. Bainbridge of Harvard, Nier built an instrument that had higher resolution and would allow measurements on lead and uranium. He then studied "the isotopic composition of Hg, Xe, Kr, Be, I, As, and Cs." Later, at the end of the 1930s, Delta Magnets to focus electron beams came into being. If the apex of the field of the Delta Magnet and the particle collector of the charged particles fell on a straight line, then the diverging beam of charged particles would focus at the collector. Nier and others reduced the size of the electromagnet used.[19]

In the Manhattan Project these mass spectrometers made uranium and hydrogen isotope evaluations in different separation plants, Monitored process steam composition in gas diffusion plants to separated uranium isotopes, And checked the vacuums of these plants with helium leak-detector mass spectrometers.[20]

Logic, Dialectic and Physics

## John Tate:  Physicist--A Profile
Early Years

Tate's father was a doctor of Scottish background.  He was a son of the American Revolution.  His mother was Irish in background.  She died when he was ten.  After the death, Tate went to New York City to live with his uncle's family.  There, he went to the Horace Mann School.  Tate went to Nebraska to College because he could be by his father who was a physician on the Rosebud Indian Reservation.  In 1912 Tate published an article entitled, "The Theoretical and Experimental Determination of Absorbing Media," in the Physical Review. His dissertation was on "The Heat of Vaporization of Metals."  Also, Tate's early efforts supported the existence of energy levels.  He made a distinction between the energy needed to instigate radiation and the energy required to create ionization.  This research showed that the quantum concept was plausible.  Heisenberg, Schrodinger and other scientists developed quantum mechanics later.[1]

During the 1920s Tate's students did research on the impact of electrons on gases.  For the years 1929-1930 E. C. Condon, a professor at Minnesota, did work on molecular binding.  He used quantum mechanics to predict that within a molecule or molecular ion there could be both an attraction as well as a repulsive potential energy.  Thus, a molecule stimulated by electrons could reach a repulsive state and form an atom and an ion.  Both particles would have measurable kinetic energy.  Bleankney saw the atomic hydrogen ions created in the dissociation.[2]

Tate did work on the ionization of molecules with electron.  Also, under Tate's purview The Physical Review became the world's leading physics journal.  He was a leader in the "quantum-mechanical revolution."  Another interest of Tate's and D. M. Dennison's was on molecules--how they rotate.[3]

Tate also maintained that science was a critical element in a liberal education.  He played golf, bridge, tennis, pool and bridge.  After Tate's first wife, Lois Beatrice Fossler who he married on December 28, 1917, died in 1939, he had some lonesome years.  Later, on June 30, 1945, he married Madeline Margarite Mitchell his office manager.  She was the Publications Manager of the American Institute of Physics.[4]

### Finis

After Tate had a stroke during December of 1949, he decreased his work load.  However, he continued working and Tate attended a meeting of the

**Logic, Dialectic and Physics**

American Philosophical Society. He died of a cerebral hemorrhage on May 27, 1950. Tate served as Editor of the American Physical Society for nearly twenty-five years. He passed away before a jubilee issue of the <u>Reviews of Modern Physics</u> dedicated to him came out.[5]

Logic, Dialectic and Physics

### Dedication of the Tate Laboratory of Physics
at the University of Minnesota:  June 21, 1966
A Thesis:  Tate Was an Alpha Male

William G. Shepherd addressed the American Physical Society who attended the dedication.  He presented a brief biography of the late professor John Tate.  On July 28, 1889 Doctors delivered John Tate in Lennox, Iowa. He got his B.A. from the University of Nebraska during 1910 and later an M.A. from there.  In 1914 he received a Doctorate of Philosophy from the University of Berlin.  After teaching at the Nebraska for two years, he moved to the University of Minnesota during 1916.  There, he became a full professor in 1919.  In 1937 Tate became a Dean at the University of Minnesota.  He held this position until he went on leave to go to war.[1]

Tate served his country with great honor in WWII as head of Division 6 of the National Defense Research Committee.  He received the President's Medal of Merit and the British King's Medal for his efforts.  Dr. Tate was the Editor of the American Physical Society from 1926 to his death on May 27, 1950.  His contribution to physics was enormous.[2]

William Shepherd then on behalf of the University of Minnesota's Regents named the building the Tate Laboratory of Physics.  Shepherd then recalled November 30, 1928, when they dedicated the original building, the American Physical Society was   present.   Professor Henry A. Erikson remembered this ceremony in his diary.  Erikson said that now the physics department had four divisions: "mechanics, heat, electricity and optics."  He then showed some slides.[3]

One of Dr. Tate's colleagues was J.H. Van Vleck soon a Hollis Professor of Mathematics and Natural Philosophy at Harvard University, the oldest endowed position in the United States.  Professor Van Vleck became a member of the faculty at Minnesota in 1923, and became a full professor in 1927-1928.  Another colleague of Professor Tate's was Professor E.U. Condon who was a professor of theoretical physics at Minnesota during 1929-1930.  A third student of Professor Tate's was Dr. A.O. Nier.  Dr. Nier graded at one time the papers of Professor Shepherd."[4]

Nier became a teaching assistant in 1933 in the physics department at the U. of M.  After getting a Ph.d, he became a National Research Council Fellow at Harvard during 1939-1938.  Later, he joined the physics faculty at Minnesota in 1938.  Nier became the department Chair from 1953-1965. Then, he became a Regent's Professor at the University of Minnesota.[5]

Logic, Dialectic and Physics

John Tate's son could not be at the dedication. He sent a letter thanking the participants for "this magnificent tribute to my father." The son wrote from Paris where he had prior commitments.[6]

## J.H. Van Vleck's Tribute

Van Vleck mentions that Minnesota is the largest western University. He started his graduate teaching career at Minnesota and worked for John Tate there. It was at Minnesota that Van Vleck met his wife, Abigail Pearson Van Vleck, U. of M., 1925. He remembers when the North Star Limited and the M. and St. L.went through the campus. The tracks went right under the present area of Northrup Auditorium. Later, developers removed these tracks. [7]

Van Vleck notes that a major physics discovery occurred during 1918 and 1923. Joseph Valasek discovered ferro-electricity. Valasek also studied the piezoelectric effect. Van Vleck remembers that appropriations for books were only about $100/yr. Also, three very promising students from Walla, Walla Washington's Whitman College came to Minnesota. They were Brattain, Bleakney, and Rojansky. Bleakney worked at the Glacier Park Hotel Company.[8]

What was, in Van Vleck's estimation, Tate's impact in the department? He was the guiding spirit of research here for a third of a century. Tate was generous to his graduate students and a fine administrator. For example, when Tate drove to gatherings of the Physical Society in Chicago or Washington, he often took students with him for a free ride. Tate was loyal to Minnesota. He turned down many offers to leave. Moreover, Tate became the Editor-in-Chief of The Physical Review.[9]

Van Vleck remembers E. Schrodinger's visit to the U. of M. He gave Schrodinger two books of Mark Twain. Van Vleck mentions that Quantum Mechanics broke during 1923-1928. Following this, Tate learned Quantum Mechanics. The total, Van Vleck claims, of distinguished Minnesota physicists was the highest of any university. Some of Tate's students took his course more than once. Peers called him a genius.[10]

## Edward U. Condon's Tribute

Condon first contacted Tate through a letter to The Physical Review. Tate published his essay on the Franck-Condon principle which explained "intensity transitions in molecules." Condon claims that Tate was a modest person with a fine sense of humor.[11]

## Logic, Dialectic and Physics

One of Tate's students was Walker Bleakney who studied ionization in mercury atoms. He made a mass spectroscopic evaluation and measured electron loss in atoms. The electron had a part in "forming a bond." After a talk with Tate, Bleakney did some more research and found "ions had some kinetic energy." Tate also worked with Philip Smith to measure ionization mass sections. Condon points out that it was Tate who conceived the idea of creating the American Institute of Physics in New York during the 1930s.[12]

### Alfred O.C. Nier's Reminiscences of John Tate

Nier remembers that Tate joined the staff in 1916. Professor Erikson sent John Zeleny a letter that Tate accepted an offer to be an instructor for $1500. By 1920 when he was 31, Tate became a full professor. In the early years Tate taught mechanics. Later, he taught theoretical physics. In 1937 Tate was chosen to be the Dean of the College. In the 1920s he taught a "Seminar in Contemporary Experimental Physics." Nier claims that Tate's lectures brought his students to the "forefront of knowledge." Tate was the advisor for over half of the 48 students who got doctorates.[13]

Nier points out that it was Tate who in 1933 helped John Williams develop an agenda in nuclear physics. Williams and others asked the Rockefeller Foundation for funds to build the Van de Graff generator outside the physics structure. Tate helped Nier get a job at the U. of M. and to get funds to buy a magnet to start his research. It was Tate, Nier says, who began the cosmic ray program. It made Minnesota a leader in this type of research. In 1939 Tate became the President of the American Physical Society.[14]

What did Tate do during WWII? He was the head of Division 6 of the National Defense Research Committee created by Vaneevar Bush. This group was instrumental in driving the German U-boats out of the North Atlantic convoy lanes. They created the anti-submarine detectors and anti-submarine devices. Tate became in 1943 the technical head of the 10th Fleet research division under Admiral E. King. An outcome of this research was the creation of the Office of Naval Research.[15]

In 1944 the Regents of the U. of M. gave Tate a special title: Research Professor of Physics. After the war on January 1, 1946, Tate returned to the physics department where he taught quantum mechanics and "atomic and molecular structure." When Tate died in 1950 he was the head of the Committee on Undersea Warfare of the National Research Council.[16]

Logic, Dialectic and Physics

## Another History Perspective

## From a History of the University of Minnesota

Frederick Jones studied with Helmholtz in Bonn, in Berlin with Kundt, and in Zurich with Weber. The department did not guarantee his position upon his return. However, Jones came back to Minnesota in 1889 to lead the physics department towards twenty years of sustained development.[1]

John Zeleny in 1892 came back to the University of Minnesota after teaching high school for a while. Both Jones and Zeleny eventually went to Yale. However, Zeleny worked for twenty-five years at Minnesota before leaving for Yale. He studied the velocities of ions created in gases by Roentgen rays. During his stay at Minnesota Zeleny invented two electrical instruments. One was to gauge the temperature of stored grain. The second instrument estimated the amount of corn in a sample of corn. These inventions helped the agricultural community. Zeleny also analyzed the movement of soil by wind.[2]

During October 1935 the U. of M.'s regents created the Institute of Technology. In so doing, they united Engineering, Architecture, Mines and also Chemistry. Physics and geology remained in the Arts College.[3]

Tate brought The Physical Review to Minnesota. He developed three major physicists: J.W. Buchta, Alfred O.C. Nier, and John H. Williams. During March and April 1940, Nier found U235 to be responsible for fission in uranium. Following this, came the separation of uranium for the atomic bomb.[4]

Williams, a University of California Ph.d, came to Minnesota to work with Tate on nuclear physics. Williams worked first with a x-ray machine, later with a Van de Graaf atom smasher, and finally with linac. During WWII most of the Minnesota physicists worked on the atomic bomb. Nier worked in New York and at Oak Ridge. Williams went to Los Alamos to finish the bomb. Nier kept the bomb project on schedule. Nier used his mass spectrometer to weigh atoms. As an aside, Tate worked together with an education psychologist, with Thomas R. McConnell and Coffman. They built up the significance of education, the humanities and the social sciences.[5]

Logic, Dialectic and Physics

## A Nuclear Particle Accelerator

### News Release

A 50 million-volt proton linear accelerator, the world's strongest "atom smasher," will be built at the University of Minnesota with a $728,000 grant given by the Atomic Energy Commission, President J.C. Morrill revealed on July 20, 1949. Dr. John H. Williams led the project. It is a research device and will not create atomic bombs. The accelerator will move hydrogen ions-a hydrogen atom without its electrons-to high speeds. These ions will disintegrate and interact with the nuclei of different atoms to help us explain the forces that bind these nuclei together.[1]

This machine resembles a large gun with a 100 foot barrel. Its diameter is five feet. The linac will send ion bullets into an atom at 60,000 miles-per-second. Scientists will load ions in the end of the "gun" which will boost them with 500,000 electron volts. This gun is half-million volt atom smasher which will power an electron transformer. An ion has a positive charge. It is to this charge that a force applies an electrical field to speed up the ion.[2]

These ions enter cavities where they will gain about 625,000 electron volts to speed them up to 30,000 miles-per-second through the accelerator. The linac uses for its power high-frequency radar oscillators on the outside of linac. After ions have moved to 60,000 miles-per-second for 100 feet, they are thrown at targets composed of atoms.[3]

A vacuum container holds these targets. The speeded up ions will break the nuclei of the target atoms like a "rifle bullet will explode an egg." Hydrogen atoms were the first targets. Later, the scientists targeted more complex atoms. The ions "bounce off target nuclei which they fail to smash." At the time, this accelerator was the world's most powerful stated Dr. Williams. Another type of atom smasher, the cyclotron, had a circular track. At Berkeley the cyclotron, at this time, had 350 million volts. It had seven times as many volts as the Minnesota accelerator had.[4]

The U. of M.'s Van de Graaf generator (1938) had about four million volts. Scientists fashioned the more powerful linac after the Berkeley linear accelerator. Professors, Williams, Henry E. Hartig, John Tate, William G. Shepherd, Charles L. Critchfield among others worked on the design of the linac.[5]

Logic, Dialectic and Physics

## A New Laboratory at Minnesota
### The Lab as Idea
The University of Minnesota's Physics Laboratory: 1940

The physics building can meet the needs of nearly 1000 undergraduates and 25 graduate students engaged in research. There is room for 9 full-time staff members. On the first floor there are three lecture rooms. They have capacities of 500, 200, and 165 seats. These lecture rooms have fixed tables for lecturing. The tables have gas, air and electric outlets. Single phase and three-phase outlets for electricity are present. These outlets can handle 110-220 volt d.c., a.c. currents. Lighting regulates from the tables as well as near the entrances to the lecture halls.[1]

White plaster projector screens are at the front and side walls in the lecture rooms. The speaker can control the lanterns from the table. Using the projector, a class can see the images of an electroscope. Galvanometers project to the walls. The lecture table has a sink with hot and cold faucets, and the floor has a drain to dissipate spilled water. Lights illuminate the blackboards. Also, the lecture table has ceiling lights.[2]

### Laboratories

The undergraduate laboratories can hold 12 to 16 students in separate rooms. Each section is run by a graduate assistant. For mechanics, heat, light and electricity, the lab rooms come in groups of three. Students would study individually and in groups. The students change rooms for different experiments. The sections handle around 144 students at once. The optics laboratories contain windows that have doors that can darken the lab. The laboratories contain gas, air and electricity outlets in the walls and tables. There are larger "Modern Experimental Laboratories" in the basement.[3]

### Research Capabilities

Most of the research facilities can handle several projects. These research rooms have brick piers which are table size. In the walls of the labs there are channel irons which can hold bolts. These channel irons can support heavy equipment or shelves.[4]

Logic, Dialectic and Physics

## Electric Facilities

In the basement there are "5 motor generator sets: 5kw, 110v; 15kw, 110v. (These are driven by synchronous motors.): 5kw, 250v; 15v, 100 amp, for changing storage batteries...and 3.5kw, 750v, for changing high voltage batteries." They are all controlled by a switchboard. This switchboard is largely a group of "bus-bars" that surround the room.[5]

The standard electric outlets exist throughout the whole building. Powerful current circuits, for example, a 200 amp size reach several of the labs. The basement batteries have 110 cells (220) v high ampere-hour capability and 3000 cells (6000v) low ampere-hour capability.[6]

## Shops

In the basement there are a wood shop and a metalworking shop. The University of Minnesota's instruments shops occupy three rooms in the basement. Each of these shops has a store room for supplies such as metal, wood or glass.[7]

## Heating

The offices and lecture rooms have individual heating systems. The building's ventilating system divide into sections that have there own sub-basement fan.[8] The heating is adequate.

## Features

The interior walls house a space of 2 feet between them to hold conduits, water pipes and ventilators. The first floor has display cabinets which contain experiments for public use. In the sub-basement there are two rooms for old equipment. The building contains a Van de Graff generator. The control room for this generator is in the basement. There is an astronomy dome on the roof. It was largely Professor Henry A. Erikson who planned the building. This was the second physics building that he helped design.[9]

## Conclusions

The key to the history of the physics department at the University of Minnesota is dialectics in the sense of "growth." The research, professors and students presented a new world view. Jones, Zeleny, Erikson, Buchta, Van Vleck, Tate, Nier and others captured the spirit of physics. Their research and

**Logic, Dialectic and Physics**

existence are a product of logic and dialectic. Is there a way of linking up the history of the physics department at the University of Minnesota with Hegel? Yes, there is. Errol Harris mentioned the Anthropic Principle or principle of man in his analysis of Hegel. A presentation of this approach follows.

Logic, Dialectic and Physics

## A Principle of Man

The Microcosm is the Macrocosm.  Patrick O'Dougherty
"All science is cosmology." K. Popper[10]

"Weak Anthropic Principle (WAP):  The seen values of all physical and cosmological quantities are not equally probable but they take on values restricted by the requirement there exist sites where carbon-based life can evolve.  Also is the requirement that the universe be old enough for it to have already done so."[11]

Strong Anthropic Principle (SAP):  The universe must have those properties which allow life to develop within it at some stage in its history. "The laws of nature must can produce life."  A corollary to this is the design argument which states "there exists one possible universe 'designed' with the goal of generating and sustaining 'observers.'"  This interpretation is religious. "Observers are necessary to bring the universe into being."  "Final Anthropic Principle (FAP): Intelligent information-processing must come into existence in the Universe, and once it comes into existence, it will never die out."[12]  This is an article of faith.

## The Design Argument and Cosmology

The cosmological argument is a proof for God's existence.  It contends that something exists and there must be reasons for all that exists.  The reasons for existence fall outside the universe.  There is a hierarchy of causes. Thus, the universe must be rational.  The mind organizes the structure of the universe. People perceive this.  This is the Anthropic position.  Objects exist and contingent articles must also exist.  Something exists, therefore, there must be a reason for its existence--God.[13]

Is there any proof for idealism?  Some computer programmers see the universe as a program run on an abstract computer.  Thought composes reality. The universe is like its simulation on a computer.  The intellect is also similar to a computer program.  Hegel would argue that a "Universal program" determines much of the content of the subprograms.[14]

Hermann von Helmholtz argued that entropy means that the universe is consuming its available energy and life would die at maximum entropy.  The universe is progressing to a lower state and is not teleological; instead, it is dysteleological.  The universe will have a "heat death."[15]

104

## Logic, Dialectic and Physics

Teilhard de Chardin accepts the concept that God is evolving. He combines Catholicism with evolution.[16] What is God's consciousness? It is unceasing prayer.[17]

A.S. Eddington in his 'Fundamental Theory' tried to predict by combinatorics the interaction strengths and particle masses. He tried to deduce by logic the quantitative propositions of physics. He suggested that the number of particles in the universe might play a role in determining the constants. For example, "the gravitational constant, the velocity of light, the electron and proton masses, the electron charge and Planck's constant" relate to the number of particles in the universe. His work met with skepticism from the scientific community. It does depict a unity in nature. The large number problem in physics remains unsolved.[18]

Paul Dirac in his Large Number Hypothesis predicted that gravity would weaken with the passage of time in the cosmos.[19] Thus, Eddington's large number would vary with time. This would qualify his thesis.[20]

If the universe is a ball then its expansion rate Ho, is Hubble's constant.[21] The universe is "inflationary."[22] It is expanding like a soap bubble.[23] A model quantum universe where the wave function is not directly dependent on the time parameter would allow a universe to start "out of nothing."[24]

### Man and Quantum Mechanics

In the Newtonian universe man was a cog. In Quantum Mechanics man has a higher stature. Man's perception changes reality. There may be a branching of the quantum universe. There may be a plurality of universes.[25]

Fred Hoyle and Richard Gott depict a 'bubble universe' where the visible portion is only one of many bubbles in a chaotic universe. In an inflationary universe the bubble universe has properties because of the way it condenses.[26] Fred Hoyle notes the 'coincidences' that occur between numerical values in the constraints of nature show a design "scheme."[27] The steady state model provides for "the continuous creation of matter." The inflationary model links together several independent properties in the universe that create life. It is a causally disjointed universe.[28]

White dwarfs become black dwarfs. Proton decay will doom man whose life relates to protons. Mass converts into energy by this process. Black holes will radiate away their mass in 10 to the 66th power more years.[29]

105

### Logic, Dialectic and Physics

A.S. Eddington thinks that entropy is the supreme law of nature. An intelligent person is like a computer with its strengths and limitations. It is the soul of a scientist. Only in a closed universe is it possible for all time based curves to be in contact with each other.[30]

If life evolves in all universes in a quantum system of cosmology then eventually an Omega Point will emerge where life will gain control of "all matter and forces not only in a single universe, but in all universes whose existence is logically possible." An infinite amount of information will emerge, "including all bits of knowledge which it is logically possible to know. And this is the end." This will occur in a closed universe.[31]

In a closed universe there is an "initial singularity, then life begins on earth, then man colonizes space , then man exhausts the earth's resources. Then, the universe will contract, sheer energy will become the energy source. The exhausted regions will re-colonize, then life will engulf the entire closed universe.[32] The physics department at the University of Minnesota is a step or microcosm in this process.

Logic, Dialectic and Physics

## Why Hegel and Physics?

The writer is a practicing Catholic, but, Hegel attracts him because Hegel hits at a loftier level, the Spirit and the Absolute, than many physicists tackle.[33] There are polarities in a science like physics. There are many grey areas in physics like the graviton. The inverse is a common mathematical operation, like the dialectic of matter and antimatter. Computer programs can sort out the equations behind the universe. Computer bits can represent nonsentential truth. Fibonnaci numbers show a design in nature. Many truths in physics are partial truths in contrast to religious truths. Is evolution converging towards an absolute like death? Perhaps it is. Physics has a unity. Physics' truths are a process. Sometimes inductive truth is the inverse of deductive truth.

Is God like a programmer? Perhaps, he is. Hegel's philosophy has limits. Partial truths are his point. A baby has fear of falling. Is that an a priori truth? Perhaps it is. Realism does not explain cognitive development very well. There are levels of abstraction. Is God beyond abstraction? Christ is not. There are two sexes in nature--a dialectic. Some languages, like French, are more dialectical than others.[34] **In contrast to Barrow and Tippler's thinking, maybe man is the anti-absolute. He is mortal.** Are there linear dissipative structures that are becoming more organized? Barrow and Tippler think there are. Physics' ideas stand for "growth, metamorphosis," deviation, and logical processes like the dialectic. Gravity is a "concrete universal." This is a paradox. Accidents and insight play a role in physics truth but so do discrepancies and dialectics. Exchange particles lie in the realm between matter and energy. Matter and energy are ideas. How did they come out of the big bang? Maybe they were there initially and ideas are the basis of the universe. "In the beginning was the word." Logic and language are an organizing principle of nature.

**"There are three kinds of people: some people think about things, some about people, and some about ideas."[35] The mathematical properties of the commutative, associative, and distributive support Hegel's idea of the triadic nature of truth. The antithesis or the negation of the dialectic are like the inverse. Hegel hits at a higher plane than the realists. A materialistic conception of the universe is slavery or what James O'Dougherty calls "manure philosophy."[36] Logic is the basis of the computer. Logic is the basis of physics and the universe. Logic is the foundation of research at the physics department at the University of Minnesota. What is an existential approach to Hegel and dialectics? Mike Franey thinks that "existence is logic."[37] The**

Logic, Dialectic and Physics

ancient Greeks invented dialectics and Aristotle formalized logic and physics. In this book Hegel's logic, specifically the dialectic, reinvents the field of physics.

Logic, Dialectic and Physics

## Bibliography

Asimov, Isaac. <u>The History of Physics</u>. New York: Walker and Co., 1966.

Barrow, John. Tipler, Frank. <u>The Anthropic Cosmological Principle</u>. New York: Oxford University Press, 1986.

Bates, Julie. A friend.

Buchta, J.W., "The Physics Laboratory at the University of Minnesota," <u>American Journal of Physics</u>, Vol. 8, No. 6, December 1940, 375-376.

Capra, Frifjof. <u>The Tao of Physics</u>. New York: Bantam Books, Inc., 1980.

Couch, William T. and Crawford, David., eds., s.v., "Hegel, Georg Wilhelm Friedrich," <u>Colliers Encyclopedia</u>. New York: P.F. Collier & Son Corporation, 1959., Vol. 9, 616-619.

Erikson, Henry A. <u>Retrospect</u>. University of Minnesota Archives, Walter Library, Minneapolis, 1941, 1.

Farber, Eduard, "Hegel's Philosophy of Physics," <u>Journal of the Washington Academy of Sciences</u>, December, 1966, 218.

Franey, Dr. Michael. A friend.

Gamow, George. <u>The Biography of Physics</u>. New York: Harper & Brothers, Publishers, 1961.

Gray, James. <u>The University of Minnesota, 1851-1951</u>. Minneapolis: The University of Minnesota, 1951.

Grier, Philip T. editor. <u>Dialectic and Contemporary Science</u>. New York: University Press of America, 1989.

Guralnik, David B., editor, s.v., "Contrapositive," <u>Webster's New World Dictionary of the American Language</u>. New York: Simon & Schuster, 1980., 309.

Hellebuyck, Roger. A friend.

Joad, C.E.M. <u>Guide to Philosophy</u>. Oxford: Oxford University Press, 1935.

Logic, Dialectic and Physics

Kolman, Bernard and Shapiro, Arnold. "Principle of Inverse Variation," Algebra for College Students. New York: Harcourt Brace Jovanovich, Publishers, 1986.

Kuhn, Thomas S. The Structure of Scientific Revolutions. Chicago: University of Chicago Press, 1970.

Nier, Alfred, "Some Reminiscences of Mass Spectrometry and the Manhattan Project," Journal of Chemical Education. Vol. 66, May 1989, 385.

----,"Some Reflections on the Early Days of Mass Spectrometry at the University of Minnesota," International Journal of Mass Spectrometry and Ion Processes, 100 (1990), 1.

----,and Vleck, John H. Van, "John Torrence Tate, 1889-1950," Reprinted from Biographical Memoirs, Vol. XLVII, (Washington, D.C.: The National Academy of Sciences of the United States, 1975, University of Minnesota Archives, Walter Library, Minneapolis), 461-465.

Noble, David. Department of History, University of Minnesota.

O'Dougherty, James. My father.

O'Dougherty, John. An uncle.

O'Dougherty, Margaret. A sister.

O'Dougherty, Terence. An uncle.

Reingold, Nathan and Reingold Ida H. editors. Science in America: A Documentary History, 1900-1932. Chicago: The University of Chicago Press, 1981.

Sambursky, Shmuel. Physical Thought from the PreSocratic to the Quantum Physicists. New York: Pica Press, 1974.

Scholz, Heinrich. A Concise History of Logic. New York: Philosophical Library, Inc., 1961.

Solomon, Robert C. In the Spirit of Hegel. Oxford, England: Oxford University Press, 1983.

Logic, Dialectic and Physics

Tate Laboratory of Physics. Dedication of the Tate Laboratory of Physics at the University of Minnesota, June 21, 1966, University of Minnesota Archives, Minneapolis, Walter Library, 1-2.

Tate, John, T. "University of Minnesota Receives $728,000 Grant from Atomic Energy Commission to Build, Operate 50 Million-Volt 'Atom Smasher,'" University of Minnesota News Service, July 19, 1949, (University of Minnesota Archives, Walter Library), 1.

Thagard, Paul, "Hegel, Science and Set Theory," Erkenntnis, 18 (1982), 397-398.

Wagner, Sister Mary Anthony. College of St. Benedict. St. Joseph Minnesota.

Wallace, William. The Logic of Hegel. Oxford: Clarendon Press, 1892.

Wolf, Abraham. Textbook of Logic. New York: Crowell-Collier Publishing Co., Collier Books, 1962.

Wright, Franklin D. and New, Bill D. "Properties and Rules," Introductory Algebra. Dubuque, IA: Wm. C. Brown Publishers, 1990.

Logic, Dialectic and Physics

**Logic, Dialectic and Physics**

**Logic, Dialectic and Physics**

Logic, Dialectic and Physics

Logic, Dialectic and Physics

Logic, Dialectic and Physics

Yang  6, 54, 59, 60, 62, 64
Yin  6, 59, 60, 64
Young  7, 23, 36-38, 76
Yukawa  54, 62
Zeleny  73, 80-83, 86-89, 98, 99, 102
Zen  60, 64
Zen Koan  64
Zeno  26
Zworykin  46

## Footnotes

### Introduction

1. Dr. David Noble, Department of History, University of Minnesota, in a private conversation.
2. D. Franklin Wright and Bill D. New, "Properties and Rules," Introductory Algebra (Dubuque, IA: Wm. C. Brown Publishers, 1990), 40.
3. Bernard Kolman and Arnold Shapiro, "Principle of Inverse Variation," Algebra for College Students (New York: Harcourt Brace Jovanovich, Publishers, 1986), 203.

### Footnotes to C. Joad's Philosophy

4. C.E.M. Joad, Guide To Philosophy (Oxford: Oxford University Press, 1935), cover.
5. Ibid., 402-404.
6. Ibid.
7. Ibid., 404.
8. Ibid., 405-410.
9. Ibid., 410-412.
10. Ibid.
11. Ibid., 413-424.
12. Ibid., 465-473.
13. Ibid., 470-477.
14. Ibid., 494.
15. Ibid., 528-529.

### Footnotes to Hegel and Physics

1. Eduard Farber, "Hegel's Philosophy of Physics," Journal of the Washington Academy of Sciences, December, 1966, 218.

**Logic, Dialectic and Physics**

2.    Dr. Michael Franey, a friend, in a private conversation.
3.    Farber, op. cit., 218-219.
4.    Ibid., 219.
5.    Ibid., 220-221.
6.    Dr. Michael Franey, a friend, in a private conversation.
7.    Farber, op. cit., 222.
8.    Ibid., 223-224.
9.    Ibid., 224-225.
10.   Ibid., 224-225.

### Footnotes to Paul Thagard's Essay

1.    Paul Thagard, "Hegel, Science and Set Theory," Erkenntnis, 18(1982), 397-398.
2.    Ibid., 398.  Robert Solomon uses the word "growth" to mean dialectics.
3.    Ibid., 398-400.
4.    Ibid., 405.
5.    Ibid., 406-407.
6.    Ibid., 408.

### Footnotes on Dialectics and Science

1.    Philip T. Grier, editor, Dialectic and Contemporary Science (New York: University Press of America, 1989), Introduction.
2.    Ibid.
3.    Ibid.
4.    Ibid.
5.    Ibid.
6.    Grier, editor, Dialectic and Contemporary Science, 3.
7.    Ibid., 4-6.
8.    Ibid., 7-11.
9.    Ibid., 12-16.
10.   Ibid., 16-24.
11.   Ibid., 24.
12.   Ibid., 25.
13.   Ibid., 26.
14.   Grier, editor, Dialectic and Contemporary Science, 29-34.
15.   Ibid., 36-41.
16.   Ibid., 36.
17.   Ibid., 53-54.
18.   Grier, editor, Dialectic and Contemporary Science, 56-57.
19.   Ibid., 57-60.

## Logic, Dialectic and Physics

20.  Ibid., 61-63.
21.  Grier, editor, Dialectic and Contemporary Science, 69-71.
22.  Ibid., 72-74.
23.  Grier, editor, "The Identity of Thought and Being in Harris's Interpretation of Hegel's Logic," op. cit., 80-85.
24.  Grier, editor, Dialectic and Contemporary Science, 91-107.

## A Portrait of Hegel

1.  William T. Couch and David Crawford, editors, "Hegel, Georg Wilhelm Friedrich," Colliers Encyclopedia (New York:  P.F. Collier & Son Corporation, 1959), Vol. 9, 616-619.
2.  Ibid.
3.  Ibid.
4.  Ibid.
5.  Ibid., 617-618.
6.  Ibid., 618.
7.  Ibid.
8.  Dr. David Noble, Department of History, University of Minnesota.
9.  See John Barrow and Frank Tipler's, The Anthropic Cosmological Principle (New York: Oxford University Press, 1986).

## William Wallace:  The Logic of Hegel

1.  William Wallace, The Logic of Hegel (Oxford:  Clarendon Press, 1892), 30-36.
2.  Ibid., 36-52.
3.  Ibid., 49-59.
4.  Ibid., 76-82.
5.  Ibid., 82-97.
6.  Ibid., 82-120.
7.  Ibid., 121-136.
8.  Ibid., 140.
9.  Ibid., 143-190.
10.  Ibid., 197-202.
11.  Ibid., 207-213.
12.  Ibid., 215-232.
13.  Ibid., 235-282.
14.  Ibid., 287-298.
15.  Ibid., 314-317.
16.  Ibid., 323-325.
17.  Ibid., 334-343.

Logic, Dialectic and Physics

18.     Ibid., 343-379.

### A Textbook Approach to Logic

1.     Abraham Wolf, Textbook of Logic (New York: Crowell-Collier Publishing Co., Collier Books, 1962), 30-35.
2.     Ibid., 35-45.
3.     Ibid., 45.
4.     Ibid., 44-56.
5.     Ibid., 56-58.
6.     David B. Guralnik, ed., s.v., "Contrapositive," Webster's New World Dictionary of the American Language (New York: Simon & Schuster, 1980), 309.
7.     Ibid., 59-60.
8.     Ibid., 59-88.
9.     Ibid., 88-106.
10.    Ibid., 123-160.
11.    Ibid., 164-165.
12.    Ibid., 171-172.
13.    Ibid., 172.
14.    Ibid., 177-185.
15.    Ibid., 187-189.
16.    Ibid., 198-207.
17.    This quote appears in the beginning of James Joyce's, Portrait of the Artist as a Young Man.

### A History of Logic

1.     Heinrich Scholz, A Concise History of Logic (New York: Philosophical Library, Inc., 1961), 8-9. The logicians included were derived from the cover and index of this book.
2.     Ibid., 2-4.
3.     Ibid., 46.
4.     Ibid., 48.
5.     Ibid., 59.
6.     Ibid., 16.
7.     Ibid., 49.
8.     Ibid., 2.
9.     Ibid., passim.
10.    Ibid., 16.
11.    Ibid., 68.
12.    Ibid., 67.
13.    Ibid., 84.

Logic, Dialectic and Physics

14.　Ibid., 58.
15.　Ibid., 83, 134.

## A Short History of Physics

1.　Isaac Asimov, The History of Physics (New York:　Walker and Co., 1966), 1-9.
2.　Shmuel Sambursky, Physical Thought from the PreSocratics to the Quantum Physicists (New York: Pica Press, 1974), 46.
3.　Ibid., 46-47.
4.　Ibid., 47.
5.　Ibid., 50.
6.　Sambursky, op. cit., 53.
7.　George Gamow, The Biography of Physics (New York:　Harper & Brothers, Publishers, 1961), 5-14.
8.　Gamow, op. cit., 17-21.
9.　Sambursky, op. cit., 1-9.
10.　Sambursky, op. cit., 58-63.
11.　Ibid., 63-83.
12.　Ibid., 128-131.
13.　Ibid., 139-141.
14.　Ibid., 142-144.
15.　Ibid., 144-146.
16.　Sambursky, op. cit., 153-161.
17.　Asimov, op. cit.,9-36.
18.　Gamow, op. cit., 51-53.
19.　Asimov, op. cit., 9-36.
20.　Ibid., 44-52.
21.　Ibid., 69-81.
22.　Sambursky, op. cit., 176-186.
23.　Ibid., 188-192.
24.　Sambursky, op. cit., 171.
25.　Gamow, op. cit., 27-30.
26.　Sambursky, op. cit., 215-237.
27.　Gamow, op. cit., 34.
28.　Sambursky, op. cit., 238-245.
29.　Ibid., 258.
30.　Ibid., 320.
31.　Ibid., 348.
32.　Sambursky, op. cit., 355.
33.　Asimov, op. cit., 97-100.
34.　Ibid., 101-114.

Logic, Dialectic and Physics

35.  Ibid., 115-134.
36.  Gamow, op. cit., 66-67.
37.  David B. Guralnik, ed., s.v., "Bernoulli's Principle," Webster's New World Dictionary of the American Language (New York:  Simon & Schuster, 1980), 134.
38.  Gamow, op. cit., 69.
39.  Asimov, op. cit., 135-143.
40.  Ibid., 135-143.
41.  Ibid., 148-154.
42.  Dr. Michael Franey, a friend, in a private conversation.
43.  Asimov, op. cit., 154-158.
44.  Ibid., 154-161.
45.  Ibid., 162-165.
46.  Ibid., 162-165.
47.  Sambursky, op. cit., 465.
48.  Asimov, op. cit., 170-179.
49.  Sambursky, op. cit., 371.
50.  Ibid., 374.
51.  Ibid., 379.
52.  Ibid., 381.
53.  Ibid., 389.
54.  Ibid., 394.
55.  Ibid., 397.
56.  Sambursky, op. cit., 405.
57.  Asimov, op. cit., 181-192.
58.  Ibid., 181-192.
59.  Ibid., 195-197.
60.  Ibid., 197-204.
61.  Ibid., 203-208.
62.  Ibid., 208-220.
63.  Gamow, op. cit., 89.
64.  Gamow, op. cit., 93-95.
65.  Gamow, op. cit., 95.
66.  Gamow, op. cit., 95-97.
67.  Gamow, op. cit., 98-102.
68.  Gamow, op. cit., 107-115.
69.  Gamow, op. cit., 117-118.
70.  Gamow, op. cit., 119.
71.  Gamow, op. cit., 119-121.
72.  Gamow, op. cit., 121-123.
73.  Asimov, op. cit., 220-246.
74.  John O'Dougherty, an uncle, in a private conversation.

75.  Dr. Michael Franey, a friend, in a private conversation.
76.  Asimov, op. cit., 247-268.
77.  Ibid., 269-275.
78.  Ibid., 275-290.
79.  Gamow, op. cit., 69-88.
80.  Asimov, op. cit., 290-300.
81.  Ibid.
82.  Ibid., 300-305.
83.  Ibid., 305-316.
84.  Ibid., 316-325.
85.  Ibid., 326-340.
86.  Ibid., 342-343.
87.  Ibid., 348-352.
88.  Gamow, op. cit., 208.
89.  Gamow, op. cit., 173-174.
90.  Gamow, op. cit., 174.
91.  Gamow, op. cit., 185.
92.  Gamow, op. cit., 186.
93.  Gamow, op. cit., 189.
94.  Gamow, op. cit., 171.
95.  Asimov, op. cit., 30-31, 357-359.
96.  Ibid., 359-364.
97.  Ibid., 365-372.
98.  Ibid.
99.  Ibid., 372-380.
100. Ibid., 380-388.
101. Ibid.
102. Ibid.
103. Ibid., 388-396.
104. Ibid., 397-416.
105. Ibid.
106. Ibid.
107. Ibid., 417-441.
108. Ibid., 441-446.
109. Ibid., 446-459.
110. Asimov, op. cit.,460-475.
111. Ibid.
112. Gamow, op. cit., 138.
113. Gamow, op. cit., 139-150.
114. Gamow, op. cit., 151-157.
115. Asimov, op. cit., 476-479.
116. Ibid., 480-498.

Logic, Dialectic and Physics

117. Ibid.
118. Ibid., 498-502.
119. Ibid., 502-508.
120. Ibid.
121. Ibid., 508-511.
122. Ibid., 503-515.
123. Ibid., 516-520.
124. Roger Hellebuyck, a friend, in a private conversation.
125. Asimov, op. cit., 516-531.
126. Sambursky, op. cit., 461.
127. Asimov, op. cit., 516-531.
128. Ibid., 531-532.
129. Asimov, op. cit., 533-546.
130. Dr. Michael Franey, a friend, in a private conversation.
131. Asimov, op. cit., 533-546.
132. Ibid.
133. Ibid., 546-550.
134. Ibid., 551-569.
135. Ibid.
136. Ibid., 570-578.
137. Ibid., 578-581.
138. Ibid., 581-586.
139. Ibid.
140. Ibid., 581-587.
141. Ibid., 588-603.
142. Ibid.
143. Gamow, op. cit., 217-220.
144. Gamow, op. cit., 222-225.
145. Gamow, op. cit., 229.
146. Gamow, op. cit., 229.
147. Gamow, op. cit., 243
148. Gamow, op. cit., 261.
149. Gamow, op. cit., 267.
150. Gamow, op. cit., 267-268.
151. Gamow, op. cit., 269-270.
152. Gamow, op. cit., 272-273.
153. John O'Dougherty, an uncle, in a private conversation.
154. Gamow, op. cit., 273-274.
155. Gamow, op. cit., 274.
156. Gamow, op. cit., 284.
157. Asimov, op. cit., 604-624.
158. Ibid.

159. Ibid.
160. Ibid., 625-632.
161. Ibid., 632-642.
162. Ibid.
163. Ibid., 643-651.
164. Ibid., 651-659.
165. Ibid., 660-679.
166. Ibid., 679-685.
167. Ibid.
168. Ibid., 684-696.
169. Ibid., 696-736.
170. Ibid.
171. Ibid.
172. Ibid.

## Fritjof Capra: The Tao of Physics

1. Fritjof Capra, The Tao of Physics (New York: Bantam Books, Inc., 1980), 11-12.
2. Dr. David Noble, op. cit., in a private conversation.
3. James O'Dougherty, my father, deceased, in a private conversation.
4. John O'Dougherty, an uncle, in a private conversation.
5. Dr. Margaret O'Dougherty, my sister, in a private conversation.
6. Capra, op. cit., 14-19.
7. Ibid., 20-27.
8. Ibid., 23-30.
9. Ibid., 32-52.
10. Ibid., 50-52.
11. Ibid., 56-66.
12. Ibid., 67-71.
13. Ibid.
14. Ibid., 75-76.
15. Ibid., 76-78.
16. Ibid., 78-79.
17. Ibid., 79.
18. Ibid., 79-81.
19. Ibid., 83.
20. Ibid.
21. Ibid., 84-85.
22. Ibid., 85-86.
23. Ibid., 87-88.
24. Ibid., 91-92.

Logic, Dialectic and Physics

25.   Ibid., 92.
26.   Ibid.
27.   Ibid., 92-98.
28.   Ibid., 93-100.
29.   Ibid., 101-107.
30.   Ibid., 107-112.
31.   Ibid., 115-129.
32.   Ibid., 130-146.
33.   Ibid., 148-164.
34.   Ibid., 173. Julie Bates, a friend, in a private conversation, suggests that "darkness can come out of light."
35.   Ibid., 175-183.
36.   Ibid., 183-190.
37.   Ibid., 193-206.
38.   Ibid., 201-208.
39.   Gamow, op. cit., 314-321.
40.   Gamow, op. cit., 322-323.
41.   Capra, op. cit., 211-215.
42.   Ibid., 212-215.
43.   Ibid., 215.
44.   Dr. Michael Franey, a friend, in a private conversation.
45.   Ibid., 215-219.
46.   Ibid., 215-227.
47.   Ibid., 227-233.
48.   Ibid., 235-242.
49.   Ibid., 236-244.
50.   Ibid., 249-256.
51.   Ibid., 256-264.
52.   Ibid., 264-269.
53.   Ibid., 269-274.
54.   Ibid., 275-281.
55.   Capra, op. cit., 276-277.
56.   Terence O'Dougherty and Dr. David Noble, in private conversations. Dr. David Noble, op. cit.
57.   Dr. David Noble, op. cit.
58.   Dr. David Noble, op. cit., in a private conversation.
59.   James O'Dougherty, my father and David Noble, a historian.

### Solomon's Spirit of Hegel

1.    Robert C. Solomon, In the Spirit of Hegel (Oxford, England:  Oxford University Press, 1983), 3-16.

Logic, Dialectic and Physics

2.    Ibid., 16-21.
3.    Ibid., 21-29. The writer is going to use the word "growth" in Solomon's sense as a synonym for dialectics throughout this text.
4.    Ibid., 33-63.
5.    Ibid., 54-63.
6.    Ibid., 64-70.
7.    Ibid., 70-77.
8.    Ibid., 77.
9.    Ibid., 78.
10.    Ibid., 110-121.
11.    Ibid., 110-128.
12.    Ibid., 129-135.
13.    Ibid., 129-144.
14.    Ibid., 150-153.
15.    Ibid., 158-186.
16.    Ibid., 186-197.
17.    Ibid., 202-209.
18.    Ibid., 215-231.
19.    Ibid., 244-260.
20.    Ibid., 244-271.
21.    Ibid., 311-317.
22.    Ibid., 322-338.
23.    Ibid., 364-406.
24.    Ibid., 406-478.
25.    Ibid., 416-435.
26.    Ibid., 447-451.

## Retrospect

1.    Henry A. Erikson, Retrospect, University of Minnesota Archives, Walter Library, Minneapolis, 1941, 1.
2.    Ibid., 2.
3.    Nathan Reingold, Science in Nineteenth-Century America (Chicago: The University of Chicago Press, 1964), Introduction.
4.    Ibid., 59-60.
5.    Ibid., 60-61.
6.    Ibid., 62-65.
7.    Ibid., 127-128.
8.    Ibid., 129.
9.    Ibid., 200.
10.    Ibid.
11.    Ibid., 200-201.
12.    Ibid.

13.  Ibid., 201-202.
14.  Ibid.
15.  Ibid., 251-253.
16.  Ibid.
17.  Ibid., 262-264.
18.  Ibid.
19.  Ibid.
20.  Ibid., 275.
21.  Ibid., 275-276.
22.  Ibid.
23.  Ibid., 276-277.
24.  Ibid.
25.  Ibid., 277-268.
26.  Ibid., 315.
27.  Ibid., 316-317.
28.  Ibid., 323.
29.  Ibid., 323-326.
30.  Erikson, op. cit., 3.
31.  Ibid., 3-4.
32.  Ibid., 5-6.
33.  Ibid., 6-8.
34.  Ibid., 9.
35.  Ibid., 97.
36.  Ibid., 97-98.
37.  Ibid., 10-11.
38.  Ibid., 12.
39.  Ibid., 13-14.
40.  Ibid., 15-16.
41.  Nathan Reingold and Ida H. Reingold, editors, Science in America:  A Documentary History, 1900-1932 (Chicago:  The University of Chicago Press, 1981), 193.
42.  Ibid., 193-194.
43.  Ibid., 194-195.
44.  Ibid., 195.
45.  Ibid., 196.
46.  Erikson, op. cit., 17.
47.  Ibid., 18-20.
48.  Ibid., 20-22.
49.  Ibid., 23-25.
50.  Asimov, op. cit., 378.
51.  Ibid., 26-27.
52.  Ibid., 28-32.

Logic, Dialectic and Physics

53. Nathan Reingold and Ida H. Reingold, <u>Science in America, op.cit.</u>, 216-217.
54. <u>Ibid</u>.
55. <u>Ibid</u>.
56. <u>Ibid</u>., 216-218.
57. <u>Ibid</u>., 218-219.
58. <u>Ibid</u>., 219-220.
59. <u>Ibid</u>., 221-222.
60. <u>Ibid</u>.
61. <u>Ibid</u>., 346-347.
62. <u>Ibid</u>., 348-349.
63. <u>Ibid</u>., 349-350.
64. <u>Ibid</u>., 350-351.
65. <u>Ibid</u>., 433-434.
66. <u>Ibid</u>., 434.
67. <u>Ibid</u>., 434-436.
68. Erikson, <u>op. cit.</u>, 33-34.
69. <u>Ibid</u>., 35-42.
70. <u>Ibid</u>., 44-49.
71. <u>Ibid</u>., 49-50.
72. <u>Ibid</u>., 51.
73. <u>Ibid</u>., 52-54.
74. <u>Ibid</u>., 56-57.
75. <u>Ibid</u>. 58-59.
76. <u>Ibid</u>., 60-61.
77. <u>Ibid</u>., 63-64.
78. <u>Ibid</u>., 65-66.
79. <u>Ibid</u>., 67-68.

### Alfred Nier's History of the Physics Department

1. Alfred Nier in a private conversation.
2. <u>Ibid</u>.
3. Alfred O. Nier, "Some Reminiscences of Mass Spectrometry and the Manhattan Project," <u>Journal of Chemical Education</u>, Vol. 66, May 1989, 385.
4. <u>Ibid</u>.
5. <u>Ibid</u>.
6. <u>Ibid</u>.
7. <u>Ibid</u>.
8. <u>Ibid</u>., 385-386.
9. <u>Ibid</u>., 386.

Logic, Dialectic and Physics

10.   Ibid.
11.   Ibid., 387.
12.   Ibid., 387-388.
13.   Ibid., 388.
14.   Ibid., 387-388.
15.   Alfred O. Nier, "Some Reflections on the Early Days of Mass Spectrometry at the University of Minnesota," International Journal of Mass Spectrometry and Ion Processes, 100 (1990), 1.
16.   Ibid.
17.   Ibid., 2-6.
18.   Ibid., 6-9.
19.   Ibid., 9-12.
20.   Ibid., 12.

## A Tribute to John Tate

1.   Alfred O. Nier and John H. Van Vleck, "John Torrence Tate, 1889-1950," Reprinted from Biographical Memoirs, Vol. XLVII, (Washington, D.C.: The National Academy of Sciences of the United States, 1975, University of Minnesota Archives, Walter Library, Minneapolis), 461-465.
2.   Ibid., 468-469.
3.   Ibid., 468-471.
4.   Ibid., 477-479.
5.   Ibid., 479-480.

## Dedication of the Tate Laboratory of Physics

1.   Dedication of the Tate Laboratory of Physics at the University of Minnesota, June 21, 1966, University of Minnesota Archives, Minneapolis, Walter Library, pp. 1-2.
2.   Ibid., 2-3.
3.   Ibid., 3-4.
4.   Ibid., 4-5.
5.   Ibid., 6.
6.   Ibid.
7.   Ibid., 9-10.
8.   Ibid., 10-12.
9.   Ibid., 12-15.
10.   Ibid., 15-20.
11.   Ibid., 21-25.
12.   Ibid., 25-30.
13.   Ibid., 31-34.

Logic, Dialectic and Physics

14.   Ibid., 33-34.
15.   Ibid., 35-36.
16.   Ibid., 36-39.

### James Gray: The University of Minnesota, 1851-1951

1.    James Gray, The University of Minnesota, 1851-1951 (Minneapolis: The University of Minnesota Press, 1951), 106.
2.    Ibid.
3.    Ibid., 323.
4.    Ibid., 417-418.
5.    Ibid., 418-419.

### The Linac Grant

1.    John T. Tate, "University of Minnesota Receives $728,000 Grant from Atomic Energy Commission to Build, Operate 50 Million-Volt 'Atom Smasher,'" University of Minnesota News Service, July 19, 1949, (University of Minnesota Archives, Walter Library), 1.
2.    Ibid., 1-2.
3.    Ibid., 2.
4.    Ibid., 2-3.
5.    Ibid., 3.

### The Physics Laboratory

1.    J.W. Buchta, "The Physics Laboratory at the University of Minnesota," American Journal of Physics, vol. 8, No. 6, December 1940, 375-376.
2.    Ibid., 376-377.
3.    Ibid., 377.
4.    Ibid., 377-378.
5.    Ibid., 378.
6.    Ibid., 379.
7.    Ibid.
8.    Ibid., 380.
9.    Ibid., 380-381.
10.   John Barrow and Frank Tipler, The Anthropic Cosmological Principle (New York: Oxford University Press, 1986), 367.
11.   Ibid., 16.
12.   Ibid., 21-23.
13.   Ibid., 103-106.
14.   Ibid., 125-154.

15.    Ibid., 166.
16.    Ibid., 167-170.
17.    Sister Mary Anthony Wagner.
18.    John Barrow and Frank Tipler, op. cit., 224-231.
19.    Ibid., 332.
20.    Ibid., 232-234.
21.    Ibid., 373.
22.    Ibid., 440.
23.    Ibid., Frederick Hoyle, 192.
24.    Ibid., 444.
25.    Ibid., 458-494.
26.    Ibid., 192-193.
27.    Ibid., 22.
28.    Ibid., 421-436.
29.    Ibid., 648-649.
30.    Ibid., 658-675.
31.    Ibid., 675-676.
32.    Ibid.
33.    Terence O'Dougherty, my uncle, thought that religious truths like Christianity hit at a higher level than most other truths.
34.    Margaret O'Dougherty, my sister, made the suggestion about the French language.
35.    James O'Dougherty, my father, in a private conversation.
36.    James O'Dougherty, my father, deceased.
37.    Dr. Michael Franey, a friend, in a private conversation.

www.ingramcontent.com/pod-product-compliance
Lightning Source LLC
Chambersburg PA
CBHW030007190526
45157CB00014B/943